Theories of Probability

AN EXAMINATION OF FOUNDATIONS

TERRENCE L. FINE

School of Electrical Engineering
Cornell University
Ithaca, New York

ACADEMIC PRESS   1973   New York and London

ACADEMIC PRESS, INC.
111 Fifth Avenue, New York, New York 10003

United Kingdom Edition published by
ACADEMIC PRESS, INC. (LONDON) LTD.
24/28 Oval Road, London NW1

LIBRARY OF CONGRESS CATALOG CARD NUMBER: 72-84364

AMS (MOS) 1970 Subject Classifications: 60A05, 62A15,
62C05, 68A20

PRINTED IN THE UNITED STATES OF AMERICA

Contents

v

III. Axiomatic Quantitative Probability

IV. Relative-Frequency and Probability

V. *Computational Complexity, Random Sequences, and Probability*

VI. *Classical Probability and Its Renaissance*

VII. Logical (Conditional) Probability

VIII. Probability as a Pragmatic Necessity: Subjective or Personal Probability

IX.   Conclusions

Preface

My interest in the foundations of probability was aroused by difficulties I encountered when first faced with applying a conventional (Kolmogorov's axioms and a limit of a relative-frequency interpretation) theory of probability to actual electrical engineering problems. Efforts to understand, usefully formulate, and resolve the problems encountered in the design and analysis of inference and decision-making systems led, it now seems inexorably, to a study of the foundations of probability. As I gradually became aware of the issues and proposals that constitute the present subject of the foundations of probability, I found myself drawn to their consideration not only for the primary pragmatic reasons but also out of respect for the breadth, depth, and provocative originality of many of the contributions to this study. It is my hope that the reader will share this appreciation.

My aim is to address all who explicitly use some theory of probability but who may not be aware of the criticisms of their preferred theory or the claims of alternative theories. Many of the difficulties encountered by engineers, physical and social scientists, and philosophers are, perhaps, attributable to misapprehensions as to the nature of the concepts of probability on which they rely. It is expected that the reader has some knowledge of a particular theory of probability. The many well-written texts on probability discouraged the inclusion of enough material to make this book self-contained as an introduction to probability. It is recommended that proofs be ignored, except by those interested in research in specific areas. To underscore this, the proofs in Chapters II, V, and VII are appended to their respective chapters, and the few proofs in the other chapters are mainly worked into the discussion.

This book can at best describe aspects of the present stage in the nascent study of theories of probability. It should raise more questions than it answers. I will be well satisfied if the critical view of probability presented stimulates the reader to the thought and research necessary to lead us all eventually to a better understanding of the work so many of us are engaged in. It is my belief that an improved understanding of the foundations of probability will induce far reaching changes in engineering and scientific practice and not merely lead to an improved justification for what we presently do.

Some years ago I set out to write a survey paper on decision theory as practiced in electrical engineering. Feeling that some of the problems of decision theory could best be understood when referred to the problems with the underlying notion of probability, I wrote an introductory section on this issue. My dissatisfaction with each draft was only temporarily allayed by an expanded redraft. This book is in fact the still unsatisfactory introduction to that as yet unwritten paper. As to its publication, I take comfort in the words attributed to Cardinal Newman that "Nothing would be done at all if a man waited till he could do it so well that no one could find fault with it."

My great debt to the many contributors to the foundations of probability, and especially to A. N. Kolmogorov, R. von Mises, and L. J. Savage, is evident throughout the book. Less evident, but no less pervasive, is the influence of discussions I have had with Messrs. Max Black, Thomas M. Cover, Zoltan Domotor, Arthur Fine, Peter C. Fishburn, Michael A. Kaplan, R. Duncan Luce, Leonard J. Savage, Herbert Shank, Georg H. von Wright, and with my wife, Susan Woodward Fine.

Introduction

IA. Motivation

Formal uses for probability and its associated concepts are found in the construction of models of random phenomena, the design of inference and decision-making systems, statements and verifications of the applicability of scientific laws, and attempts to understand knowledge and induction. Informal uses for probability include the Butlerian view of probability as a guide to life and the frequent appearance of the words "probably" and "likely" in ordinary discourse. Notwithstanding the importance of probability in the explication of knowledge and induction, its roles in the verification of laws or "as a guide to life," or its prevalence in discourse, we leave the analyses of these issues to suitably trained philosophers. Our concern is primarily with those concepts of probability that are important for the modeling of random phenomena and the design of information-processing systems.

Methods for modeling the random phenomena of chance and uncertainty and the design of inference and decision-making systems are of great importance in fields as diverse as engineering and the physical and social sciences. In electrical engineering, areas such as communications, detection, pattern classification, and stochastic control owe their very formulation to concepts of probability theory. The fundamental work

1

of Shannon [1] and Wiener [2] in electrical engineering appeared to establish probability, as supplemented by statistics, as the appropriate methodology for the modeling of noise and information sources, the determination of their characteristics, and the selection of system performance criteria. In the physical sciences probability is at the core of statistical and quantum mechanics and their manifold applications throughout science. Indeed, the very statement of quantum mechanics would not be possible without probability concepts. In the social sciences, probabilistic models and concepts are of importance to such disciplines as economics, psychology, and sociology. The language of probability is used to state models of uncertainty and time series and is basic for the statement of laws of individual and collective behavior. Furthermore, probability and statistics are used throughout engineering and the sciences for the analyses of experiments and the degree to which they confirm hypotheses; probability is an integral part of the methodology of science.

The very wide range of applications of probability and the very wide variety of random phenomena have, appropriately, induced a large number of views as to the nature of probability and its associated concepts. Yet in the physical sciences and engineering there has been a pronounced tendency to oversimplify matters and to rely on only one view of probability. While the social sciences have been somewhat more adventuresome, we do not think they have been sufficiently so. It should be clear that the multifarious and vital roles assigned to probability warrant its close examination. We need to study critically the possible formalizations (e.g., the Kolmogorov calculus) and interpretations of probability (e.g., relative-frequency). We need to increase our awareness of the diversity of random phenomena and design objectives. We need to improve our understanding of probability before we can reasonably expect to measure it. (Recourse to *ad hoc* statistical principles of estimation is both intellectually and pragmatically unsatisfactory; the prevalence of statistical legerdemain is indicative of the wide or possibly unbridgable gap between theory and practice.) We need to understand more clearly the bases for and meaning of probability conclusions before we can expect to select performance criteria, based on probability, that will enable us to design satisfactory systems. (How should you judge the reliability of a unique system? What is compelling about minimizing expected loss?)

In what follows we attempt to respond to the above challenges and lead the reader to a vantage point from which he can choose between, modify, or develop new probability-based methodologies as needed for specific applications. However, we are under no misapprehensions as to the completeness of either the criticisms we offer or the range of theories we present. Readers who are stimulated by the questions asked, and those avoided, may find some answers in the references, but they are more likely to find themselves forced to find their own answers.

IB. Types of Probability Theories

1. Characteristics of Probability Theories

We can roughly classify probability theories along the following dimensions.

(1)   The domain of application—the set of random phenomena for which it supplies a model.

(2)   The form of probability statements.

(3)   The relations between probability statements.

(4)   The information input and procedures to be used in measuring, or arriving at initial, probability statements.

(5)   The purpose or goals of the theory—the value and/or meaning of probability statements.

These dimensions are interdependent and cannot be considered in isolation. For example, relations between statements depend on the form of statements, the goals achievable by the theory depend on the method of measurement and the form of probability statements, and so on. A better understanding of these dimensions may be obtainable from illustrations of how they manifest themselves in current theories.

2. Domains of Application

At a general level, examples of domains of application of probability theories include

(1)   characterization or modeling of the occurrences of outcomes in (potentially) indefinitely repeated experiments;

(2)   explicating the degree to which statements in a language support or confirm each other;

(3)   classification of sequences, statements, or observations according to their generating mechanisms;

(4)   representation of an individual's beliefs, opinions, and preferences.

More specific examples of applications of probability theories include modeling of physical noise sources such as coin tosses, thermal noise, distribution of electrons striking a screen; modeling of information sources such as phoneme patterns in speech, fluctuations in gray levels in photographs or television pictures, digital or analog telemetry signals from a remote probe or monitor station; modeling of economic and sociological sources such as fluctuations in demand for products as reflected, for example, in time series of stock prices, demographic variables including distribution of population and work force mobility; psychological investigations of errors in perception and limits on perceptual discrimination; statements in ordinary discourse about anticipated success in ventures and the intensity of our beliefs and preferences.

3. Forms of Probability Statements

Probability statements occur in very many forms. The quantitative statement "$P(A) = p$" read "the probability of A is p" is most common in the sciences. The modal or classificatory statement "A is probable" or "A is likely" seems to be most common in ordinary discourse. One way to (partially) organize a listing of the forms of probability statements is to consider them as n-term relations P_n between elements of a domain set $\mathscr{D}$. The domain $\mathscr{D}$ might typically be a field of events or the set of propositions in a simple language. An n-term relation P_n is a subset of the set $\mathscr{D}^n$ (Cartesian product) of n-tuples of elements of $\mathscr{D}$. For different values of n we find the following kinds of probability statements.

(1)   A unary relation P_1 is a subset of $\mathscr{D}$ and can be read as follows: If $A \in P_1$, then A is probable. This is the classificatory or modal form of probability.

(2)   A binary relation P_2 is a subset of $\mathscr{D}^2$ and can be read as follows: If $(A, B) \in P_2$, then A is at least as probable as B. A more conventional notation for this form of comparative probability statement is "$A \gtrsim B$."

(3)   A ternary relation P_3 is a subset of $\mathscr{D}^3$ and, though not generally used, can be read as follows: If $(A, B, C) \in P_3$, then A given C is at least as probable as B given C.

(4)   A quaternary relation P_4 is a subset of $\mathscr{D}^4$ and can be read as follows: If $(A, B, C, D) \in P_4$ then A given B is at least as probable as C given D. A more conventional notation for this form of comparative conditional probability is "$A/B \gtrsim C/D$."

While other interpretations and higher order relations are conceivable, they are only rarely used explicitly.

To see how the quantitative form of probability statement fits into this relational scheme we need to adopt the viewpoint of the theory of measurement as treated in Section IIID. Quantitative probability becomes meaningful when viewed as a (homomorphic) numerical representation of a probability relational system. For example, a representation of P_2 is a function (quantitative probability) $P(\cdot)$ satisfying

(a) $P : \mathscr{D} \to R^1$;
(b) $(A, B) \in P_2 \Leftrightarrow P(A) \geqslant P(B)$.

Quantitative conditional probability $P(\cdot \mid \cdot)$ can be viewed as a representation of P_4 through

(a) $P : \mathscr{D}^2 \to R^1$;
(b) $(A, B, C, D) \in P_4 \Leftrightarrow P(A \mid B) \geqslant P(C \mid D)$.

There are generally many possible representations and not all are compatible with the usual quantitative theories. The quantitative forms of probability statements generally implicitly assume that P represents more than just the P_n relations. We defer consideration of this question to the later chapters on specific theories of probability.

Finally, we note that even discounting possible problems with quantitative forms, our classification of forms is incomplete. For instance, analyses of the use of "probable" or "likely" in ordinary discourse have suggested that there are many variations in usage of even this simple unary concept. Nonetheless, our classification of forms should suffice for our immediate purposes of orientation.

4. Relations between Statements

The relations between probability statements are described by a set of axioms or a calculus that enables us to derive new statements from known statements and that delineates the syntax of probability statements. The Kolmogorov axioms for quantitative probability and events comprise the best known calculus. Properties of events are delineated through the requirement that the collection of events form a σ-field (e.g., if A is an event, then so must be the complement $\bar{A}$ of A). Properties are asserted of probability statements [e.g., $1 \geqslant P(A) \geqslant 0$] and other properties enable us to derive new probability statements as well [e.g., if A contains B, then $P(A) \geqslant P(B)$]. In our examination of theories of probability we will encounter a variety of quantitative calculi as well as calculi or axioms for the modal and comparative forms of probability.

5. Measurement

Unfortunately, most theories of probability either ignore the vital issue of the measurement or initial assessment of probability or else they gloss over it (e.g., equating probability with relative-frequency or invoking raw intuition). The measurement of probability, especially in an empirical theory, requires data or an information input and a process for converting the data into probability statements. The nature of the data is largely determined by the domain of application. The measurement process converts what it recognizes as relevant data into probability statements that have the correct form and that are mutually consistent with the assumed calculus. Furthermore, the measurement process is designed so as to achieve the purposes or goals of the theory.

Examples of sources of data are probabilities themselves; frequency of occurrence of an event in unlinked repetitions of a (random) experiment; evidence statements in a simple language; data sequences that can be read by a class of machines; and judgments, opinions, and preferences distilled from individual experience. Examples of the corresponding measurement processes are nothing, or an axiomatic calculus; a presupposition that there exists a limiting frequency together with statistical principles of estimation; a logical relation of induction or confirmation that is a generalization of deductive implication; a presupposition that random phenomena are generated by certain mechanisms (probabilistic automata); and intuition, possibly assisted by a system of consistency requirements.

6. Goals and Their Achievement

Often the available descriptions of a probability theory do not lend themselves to a classification of the purposes or goals of the theory. Either the purpose is unstated or else so poorly conceived as to be unachievable by the theory or, perhaps, by any theory. Before considering the important problems of justifying whether a probability theory achieves its goals, we note the following examples of goals that have been claimed as objectives of probability theories: an assertion of a physical characteristic of an experiment that will be manifested under prescribed conditions; an elaboration of correct inductive reasoning concerned with assessments of degrees of truth; a formalization of individual opinion leading to decisions satisfactory to the individual; an expression of individual judgments in an interpersonal form; a summary description of data; and selection of a probabilistic automaton as a model of the data source.

The problem of the justification of a probability theory as achieving its goals is an intricate and tangled one. Although a theory may claim to yield probability statements of the types just indicated, it is often the case that the goals are incompatible with the first four dimensions of formulation. We can discern at least four approaches to the justification or defense of a probability theory:

(1)   internal consistency and freedom from ambiguity;
(2)   validation;
(3)   vindication;
(4)   relation to other theories.

Satisfaction of the criteria of internal consistency and freedom from ambiguity are obviously essential for the construction of a theory suited to application. We validate a theory by proving that it leads to the conclusions it claims to. A theory can be vindicated by showing that it leads to the desired conclusions if any theory does. A theory may be related to, or subsume, other theories in terms of the typology indicated above; perhaps it is applicable whenever some other theory is and in other cases as well. The nature of justification and the meanings of validation and vindication are discussed in an interesting paper by Feigl [3]. Our brief remarks on this involved subject are only intended to provide a framework for later criticism of the various theories of probability.

7. Tentative Classification of Some of the Theories to Be Discussed

The following terms also occur in descriptions of probability theories, and it may be of convenience to the reader if we informally define them here.

Empirical   factual, physical properties of the world, based solely on experiment and observation.

Logical   formalized theory of rational reasoning or correct thinking, either deductive or inductive.

Objective   interpersonal, independent of thought.

Pragmatic   emphasizing the practical and good rather than the true or correct.

Subjective   involving the thought of an individual, utilization of individual opinions, judgments, and experience.

Discussion of the meanings of these terms can be found in Carnap [4, pp. 1–51] and in texts on the philosophy of science [5].

A loose classification of some of the theories we will examine, in terms of the five dimensions we have discussed, is as follows.

Axiomatic comparative

(1), (4), (5) are ignored;
(2)   comparative probability relation;
(3)   set of axioms establishing a complete ordering of events;
        other terms are inapplicable.

Kolmogorov's calculus

(1), (4), (5) are ignored;
(2)   assignment of numbers to events;
(3)   set of axioms restricting possible probability numbers and determining the probabilities of disjoint unions of event from those of the events;
        other terms are inapplicable.

Usual relative-frequency theory

(1)   indefinitely repeatable experiments;
(2)   quantitative;
(3)   Kolmogorov's calculus;
(4)   observed relative-frequencies of occurrences and *ad hoc* statistical principles;
(5)   physical characteristic of an experiment useful for prediction; intended to be empirical and objective.

Von Mises' relative-frequency theory

(1)   indefinitely repeatable experiments of a special kind;
(2)   quantitative;
(3)   calculus of Jordan measure;
(4)   observed relative-frequencies, the hypothesis of a collective and *ad hoc* statistical principles;
(5)   physical characteristic of an experiment useful for prediction; intended to be empirical and objective.

Reichenbach–Salmon relative-frequency theory

(1)   indefinitely repeatable experiments;
(2)   quantitative;
(3)   finitely additive, unit normed, nonnegative set functions;

(4) observed relative-frequencies, rational principles of estimation;
(5) elaboration of inductive reasoning;
   logical, empirical, intended to be objective, pragmatic.

Solomonoff's complexity-based theory

(1) finite-length sequences of symbols;
(2) quantitative;
(3) uncertain, although they are intended to satisfy axioms of finite additivity, nonnegativity, and unit normalization;
(4) sequences of symbols, hypotheses as to the origins of the sequences, notions of simplicity, *ad hoc* manipulations motivated by concepts from the theory of computational complexity;
(5) summary of data, elaboration of inductive reasoning, explanations of observed sequences in terms of generating mechanisms, prediction of future terms in the sequences;
   logical, empirical, and intended to be objective.

Laplace's classical theory

(1) uncertain, applied indiscriminately;
(2) quantitative;
(3) Kolmogorov calculus or finitely additive, nonnegative set functions with unit normalization;
(4) ignorance concerning which is the most likely of a set of alternatives, no carefully worked out measurement procedures;
(5) permits probability statements in many situations, meaning uncertain;
   despite some claims, appears to be subjective and possibly logical.

Jaynes' classical theory

(1) uncertain;
(2) quantitative;
(3) Kolmogorov calculus;
(4) "testable information" and application of the principle of maximum entropy;
(5) permits probability statements in many situations, meaning uncertain;
   objective.

Koopman's comparative logical theory

(1) propositions describing experiments;
(2) comparative;

(3)   set of axioms that restrict possible comparative relations, not necessarily completely ordered;
(4)   intuition;
(5)   uncertain, possibly consistent use of judgments;
      logical, subjective.

Carnap's logical theory

(1)   statements in certain simple languages;
(2)   quantitative;
(3)   calculus of finitely additive, nonnegative set functions supplemented by axioms of invariance;
(4)   *a priori* principles of inductive reasoning;
(5)   elaboration of "rational" inductive reasoning and behavior;
      logical, objective, asserted to be pragmatic as well.

De Finetti–Savage subjective–personalistic theory

(1)   individual beliefs, judgments, and preferences;
(2)   quantitative;
(3)   axioms provide constraints of consistency or coherence;
(4)   intuition or introspection;
(5)   expression of individual beliefs in a form useful to the individual for selecting his most satisfactory decisions;
      subjective, pragmatic.

The preceding characterizations are debatable. Probability theories do not always fall neatly into our categories, and furthermore detailed formulations that would enable us to classify a theory confidently are often unavailable. A logical taxonomy of probability theories which is not merely a catalog would greatly stimulate clearer thinking about probability. In the absence of such a taxonomy our remarks may be of benefit in orienting the reader to the more extensive discussions to follow.

IC. Guide to the Discussion

1. Outline of Topics

We start in Chapters II and III by examining axiomatic formalizations of comparative and quantitative probability and the relations between them. The discussion of comparative probability emphasizes the condi-

tions that characterize those comparative probability relations representable by quantitative probability as well as the possibilities for new
definitions of important concepts such as independence. In Chapter III
we concentrate on the widely accepted Kolmogorov formalization of
quantitative probability. We conclude that the Kolmogorov formulation,
while a milestone in the history of probability and the cornerstone of
almost all modern work in probability, both lacks sufficient structure of
a meaningful kind and unnecessarily restricts the domain of application
of probability by imposing requirements of a conventional, rather than
substantive, kind. Furthermore, in the absence of an interpretation,
neither the comparative nor the quantitative theories are of any value for
practice (e.g., modeling or system design).

In Chapter IV we treat the common interpretation of probability as a
limit of the relative-frequency of the number of occurrences of an event
in repeated, unlinked trials of a random experiment. A close look at the
origins of the empirical hypothesis of convergence or stability of the
relative-frequency suggests that this hypothesis arises out of a mis-
interpretation of our experience. Nevertheless, we proceed and examine
the Bernoulli/Borel formalization of this hypothesis and its justification
through the theorems of the laws of large numbers. The inadequacy of
the Bernoulli/Borel formalization leads us to a cursory examination of the
more carefully thought out von Mises formalization of a relative-
frequency-based theory of probability. We find that neither the Bernoulli/
Borel nor the von Mises formalization lead to probability theories
that enable us to, in any practical sense, measure or use probability; they
both require supplementation by *ad hoc* principles of statistics. Other
approaches, especially an interesting one by Salmon, to relative-
frequency-based probability theories are then outlined but judged
inadequate as bases for confident application to practice.

An improved theory for repeated random experiments, related to the
ideas of von Mises, is sketched in Chapter V. Chaitin, Kolmogorov, and
Solomonoff suggested that the high irregularity and absence of a readily
discernible pattern of a random sequence could be formally characterized
through a theory of computational-complexity. While the complexity
approach greatly improves our understanding of randomness and
independence, it seems to require supplementation by debatable
assumptions before it can be developed into a theory of probability. The
construction of a complexity-based theory of probability, while doubtless
possible, is as yet incomplete.

Not all of the random phenomena we encounter in engineering can
be modeled by a relative-frequency-based probability theory, even
assuming that we have such a theory. What of chance phenomena that

are not repeatable or uncertainty about states of the world due to a lack of knowledge? Classical probability is historically the first probability theory to provide service for such problems. Chapter VI contains a description of the classical theory, as based on the use of the principle of indifference and a presumed ability to recognize the absence of a preponderance of evidence in favor of any particular experimental outcome or state of the world. Reformulations of the classical theory through principles of invariance and maximum entropy or mutual information are outlined. Unfortunately, we find the classical theory to be confused in its objectives and without sufficient guidelines for its consistent application.

A clarification of the classical approach with respect to the assessment of a balance of evidence leads to a discussion of logical probability in Chapter VII. The emphasis shifts from probability as attached to events to probability as a relation between statements; probability is a measure of the degree to which an evidence statement supports a hypothesis or the degree of uncertainty we are rationally "entitled" to have concerning a hypothesis. Logical probability is an attempt at logical induction. After briefly remarking on classificatory logical probability we then present Koopman's theory of comparative logical probability, in which assessments derive from our intuition, and indicate its relation to quantitative logical probability and relative-frequency. We then discuss Carnap's formulation of quantitative logical probability, its seemingly objective character, its relation to relative-frequency, and its claimed role as the basis for rational decision-making. We conclude that Carnap's theory falls short of its objectives either as a basis for inference or decision-making.

The introduction of questions of decision-making then directs us to consideration of the decision-oriented personalistic/subjectivistic theories of probability. Individual decision making rather than the modeling of random phenomena is the concern of subjective probability. In this respect subjective probability is distinct from the other versions of probability that we discuss. In Chapter VIII we sketch the origin of subjective probability as a by-product of the development of individual judgments, in a consistent and rational manner, into decisions, and of attempts to communicate individual judgments between people and between people and machines. The subjective element implicit in all practical applications of probability is at last made explicit by the subjective theories. However, the particular form taken by the usual subjective theories, as well as their claims to universality, are deemed open to challenge.

The conclusion reached in Chapter IX is that none of the presently known theories of probability covers the whole domain of engineering and scientific practice. Whether a useful unitary of probability is possible or whether probability is indeed necessary for the design of inference and decision-making systems are still open questions.

2. References

Our exposition provides only an introduction to some of the many concepts of probability, their formal development, and their defense. It is natural that such an exposition contain critical and speculative, as well as mathematical, comments. The study of theories cannot be conducted solely as an exercise in deductive logic. However, should our remarks prove cryptic, it is hoped that the interested or puzzled reader will avail himself of the material indicated in our references and that he will consult the broadly based discussions of, and extensive biblio-graphies on, the various theories of probability that can be found in the work of Carnap [4], Black [6], Good [7], Kneale [8], Kyburg [9], Lucas [10], Salmon [11], Savage [12], or von Wright [13]. Furthermore, although we will often attach an individual's name to a particular thesis, we must assume a responsibility for any injustice we may inadvertently do to the originators of many of the ideas presented here. The many possibilities for misunderstanding and omission lead us to recommend that when an author's name is given, the reader accept it as a reference rather than as an accurate historical remark. A modern history of proba-bility would be a major work in itself.

3. Known Omissions

The theories we discuss in this work cover a broad range but do not exhaust all that have been proposed. We treat neither all of the variants of a particular outlook nor all of the possible approaches. It is hoped that the chapter references will lead the reader to most of the variants of the types of theories we do discuss. Among the approaches that we do not effectively consider are fiducial probability, methodological approach, and ordinary language concepts of probability. We are uncertain as to the nature of Fisher's fiducial probability and expect that the discussion by Dempster [14] is sufficient to orient the interested reader. The methodological approach, mentioned in Section IXB, is usually asso-ciated with Braithwaite [15]. We do not believe that it forms a distinct theory of probability, but some might argue otherwise. We have ignored the ordinary language concepts of probability both because they are formally weak notions and because the analyses of such concepts seem best left in other hands [16].

References

1. C. Shannon and W. Weaver, *Mathematical Theory of Communication*, pp. 9–18. Urbana: Univ. of Illinois Press, 1949.
2. N. Wiener, *Extrapolation, Interpolation and Smoothing of Stationary Time Series*, pp. 1–13. New York: Wiley, 1949.
3. H. Feigl, De Principiis Non Disputandum ...? in *Philosophical Analysis* (M. Black, ed.), pp. 113–149. Englewood Cliffs: Prentice-Hall, 1963.
4. R. Carnap, *Logical Foundations of Probability*, 2nd ed. Chicago: Univ. of Chicago Press, 1962.
5. C. Lewis, *An Analysis of Knowledge and Valuation*. LaSalle: Open Court, 1946.
6. M. Black, *Margins of Precision*, pp. 57–208. Ithaca: Cornell Univ. Press, 1970.
7. I. J. Good, *Probability and the Weighing of Evidence*. London: Griffin, 1950.
8. W. Kneale, *Probability and Induction*. London and New York: Oxford Univ. Press (Clarendon), 1949.
9. H. E. Kyburg, Jr., *Probability and Inductive Logic*. New York: Macmillan, 1970.
10. J. R. Lucas, *The Concept of Probability*. London and New York: Oxford Univ. Press (Clarendon), 1970.
11. W. C. Salmon, *The Foundations of Scientific Inference*. Pittsburgh: Univ. of Pittsburgh Press, 1966.
12. L. J. Savage, *Foundations of Statistics*. New York: Wiley, 1954.
13. G. H. von Wright, *The Logical Problem of Induction*, 2nd rev. ed. New York: Barnes & Noble, 1957.
14. A. P. Dempster, On the Difficulties Inherent in Fisher's Fiducial Argument, *J. Amer. Stat. Assoc.* **59**, 56–66, 1964.
15. R. B. Braithwaite, *Scientific Explanation*, pp. 151–191, reprint. New York: Harper & Brothers, 1960.
16. S. Toulmin, *The Uses of Argument*. London and New York: Cambridge Univ. Press, 1958.

II

Axiomatic Comparative Probability

IIA. Introduction

The concept of comparative probability (CP) exemplified by the statement "*A* is at least as probable as *B*" has thus far received very little attention from either engineers, scientists, probabilists, or philosophers. Grounds for a greater interest in this neglected concept include the following points.

(1)   CP provides a more realistic model of random phenomena when we have too little prior information and data to estimate quantitative probability reasonably.

(2)   CP provides a wider class of models of random phenomena than does the usual quantitative theory.

(3)   CP illuminates the structure of quantitative probability, and especially the Kolmogorov axioms, by providing a base from which to derive quantitative probability.

(4)   CP appears to be a sufficiently rich concept to support a variety of significant applications.

With respect to (1), the logical relation between CP and quantitative probability is such that we are always more confident about estimates of the former than of the latter; for example having observed that 10 tosses of a strange coin resulted in 7 heads, we are more justified in asserting that "heads are more probable than tails" than in asserting "the probability of heads is .7." Point (2) refers to the curious phenomenon that there exist relatively simple examples of what we consider to be valid CP statements that are incompatible with any representation in the usual quantitative theory. One characteristic of these new models of random phenomena is that they do not admit of indefinitely repeated independent performances. Point (3) will, we hope, be vindicated by the discussion to follow here and in Chapter III. Point (4) is open to challenge. There has been very little work attempting to apply CP. While an application to decision making is given in Section IIG, clearly much remains to be done.

Our development of CP derives from attempts to answer the following questions.

(1) What is the formal, axiomatic structure of the relation "at least as probable as"?

(2) What is the relation between CP and quantitative probability as formalized by Kolmogorov?

(3) What are definitions of associated concepts such as comparative conditional probability, comparative independence, and expectation?

(4) How can we estimate or measure comparative probability so as to generate a useful empirical theory of probability?

(5) How can we use CP in decision-making and inference?

Questions (1) and (2) are examined in Sections IIB, IIC, and IID. We treat (3) in Sections IIE, IIF, and IIH and (5) in Section IIG. Remarks on the interpretive basis (4) are deferred to Section IVK for a relative-frequency basis, Section VI for a computational complexity basis, and Section VIIC for a logical basis. Before we can expect to resolve the measurement problem of (4) we will have to advance the state-of-the-art in decision making referred to in (5).

The results we present in this chapter are incomplete in many respects. Much more research will be required before comparative probability can take what we judge to be its highly significant rightful place in the analysis of random phenomena.

IIB. Structure of Comparative Probability

1. Fundamental Axioms That Suffice for the Finite Case

Comparative probability, hereafter abbreviated as CP, is a binary relation between events, or their representations as sets or propositions, that we denote by "$A \succsim B$" and read as "event A is at least as probable

as event B"; alternatively, we may write "$B \lesssim A$" which can be read as "event B is no more probable than event A." We define equally probable events, "$A \approx B$" to be read "event A is as probable as event B," by $A \approx B$ if $A \gtrsim B$ and $B \gtrsim A$. Finally, we define "$A > B$" to be read "event B is not as probable as event A" by $A > B$ if $A \gtrsim B$ and false $B \gtrsim A$. In the sequel the events are represented as elements of a field $\mathscr{F}$ of subsets of a set Ω (see Section IIIA) and are denoted by capital roman letters. The following five axioms partially characterize the family of CP relations $\gtrsim$ on $\mathscr{F}$.

C0. (*Nontriviality*) $\quad \Omega > \varnothing$, where $\varnothing$ is the null or empty set.

C1. (*Comparability*) $\quad A \gtrsim B$ or $B \gtrsim A$.

C2. (*Transitivity*) $\quad A \gtrsim B, B \gtrsim C \Rightarrow A \gtrsim C$.

C3. (*Improbability of impossibility*) $\quad A \gtrsim \varnothing$.

C4. (*Disjoint unions*) $\quad A \cap (B \cup C) = \varnothing \Rightarrow (B \gtrsim C \Leftrightarrow A \cup B \gtrsim A \cup C)$.

Axioms C1 and C2 establish that the CP relation $\gtrsim$ is a linear, complete, or simple order. The requirement that all events be comparable is not insignificant and has been denied by many careful students of probability including Keynes and Koopman (see Section VIIC). Although we will not explore the matter, it is possible that the concept of CP should allow for an indifference (symmetric, reflexive, intransitive) relation between events. Axioms C3 and C4 would presumably be widely assented to as essential to the characterization of "at least as probable as." These five axioms appear to suffice for the cases where $\mathscr{F}$ is a finite set. Yet as we shall see in Section IIC they admit CP orders that need not be compatible with the usual probability theory.

The consequences of these five axioms include the following easily verified reasonable properties of a CP relation.

(a) $\quad \Omega \gtrsim A$.

(b) $\quad A \gtrsim B \Rightarrow \bar{B} \gtrsim \bar{A}$.

(c) $\quad A \supseteq B \Rightarrow A \gtrsim B$.

(d) $\quad A \gtrsim B, \quad C \gtrsim D, \quad A \cap C = \varnothing \Rightarrow A \cup C \gtrsim B \cup D$.

(e) $\quad A \cup B \gtrsim C \cup D, \quad C \cap D = \varnothing \Rightarrow$ either $A \gtrsim C$,
$\qquad A \gtrsim D, B \gtrsim C$, or $B \gtrsim D$.

If $\mathscr{F}$ is required to be merely a Λ-field (see Section IIIB), then the preceding consequences of C0–C4 need not hold. In this case (d) might be an appropriate replacement for C4 [1].

2. Compatibility with Quantitative Probability

A function P is said to agree with, or represent, an order relation $\gtrsim$ if

$$P : \mathscr{F} \Rightarrow R^1, \qquad A \gtrsim B \Leftrightarrow P(A) \geq P(B).$$

When $\gtrsim$ is a CP relation, then we will refer to P as a quantitative probability. Axioms C0–C4 do not permit us to conclude that for any CP relation $\gtrsim$ satisfying C0–C4 there exists some quantitative probability.

Counterexample (Lexicographic order) [2].   Let $\Omega = [0, 1]$ be the Borel field of subsets of Ω, $l(A)$ the Lebesgue measure (length) of A, μ a measure dominated by l which for the sake of illustration we take to have the triangular density 2ω over $[0, 1]$. We now define a relation $\gtrsim$ by: $A \gtrsim B$ if either $l(A) > l(B)$ or $l(A) = l(B)$ and $\mu(A) \geq \mu(B)$. It is easily seen that $\gtrsim$ satisfies C0–C4. To verify that there is no agreeing P we consider the sets $A_x = (1 - x, 1)$ and $B = (0, x)$ where $0 < x < 1$ for any x. It is evident that $l(A_x) = l(B_x) = x$, $\mu(A_x) = x(2 - x)$, $\mu(B_x) = x^2$, and for $x < x'$ we have $B_x \prec A_x \prec B_{x'} \prec A_{x'}$. Hence, for any agreeing P we must have $(P(B_x), P(A_x))$ is a nonnull interval and $x \neq x'$ implies that $(P(B_{x'}), P(A_{x'})) \cap (P(B_x), P(A_x)) = \varnothing$. Thus we have established a one-to-one correspondence between real numbers x in $(0, 1)$ and disjoint intervals in R. However, this is a contradiction, for there are uncountably many x and countably many disjoint intervals. Therefore, there does not exist P agreeing with the lexicographic order $\geq$. ∎

In order to propose a fifth axiom to guarantee the existence of a quantitative probability we must first recall a few terms from topology [3]. A topological space $(\mathscr{F}, \mathscr{T})$ is a set $\mathscr{F}$ and a collection $\mathscr{T}$ of open subsets of $\mathscr{F}$; $\mathscr{T}$ is the topology. A base $\mathscr{B}$ for $\mathscr{T}$ is a collection of subsets of $\mathscr{F}$, $\mathscr{B} = \{B_\alpha\}$, such that $T \in \mathscr{T}$ is a union of sets in the base. A subbase for $\mathscr{T}$ is a collection of subsets of $\mathscr{F}$ such that finite intersections between elements of the subbase generate the base. A countable base is a base with a countable number of elements. The order topology corresponding to the complete, transitive, irreflexive relation $\prec$ is the topology with subbase having as elements $\{A : A \prec B\}$, $\{A : B \prec A\}$ for any $B \in \mathscr{F}$. Finally, a set $\mathscr{D}$ is order dense if

$$(\forall A, B \in \mathscr{F} - \mathscr{D}) \; A \prec B \Rightarrow (\exists D \in \mathscr{D}) \; A \prec D \prec B.$$

We may now state the following equivalent axioms.

C5a. $(\mathscr{F}, \mathscr{T})$ has a countable base.

C5b. $(\mathscr{F}, \mathscr{T})$ has a countable order dense set.

The justification for these axioms is contained in the following theorem.

Theorem 1. If $\gtrsim$ satisfies C1–C4, then it admits of a quantitative representation P if and only if $\gtrsim$ also satisfies C5, where C5 can be taken to be either C5a or C5b.

Proof. All proofs of results in this chapter can be found in the Appendix to this chapter. ∎

Theorem 1 may be of value for other axiomatizations of $\gtrsim$ when there are infinitely many equivalence classes of events ($\mathscr{F}/\approx$ is infinite); C5 is trivially satisfied when $\mathscr{F}/\approx$ is finite. A possibly more intuitive sufficient condition for the existence of agreeing P, involving an axiom of monotone continuity for $\gtrsim$, is stated in Subsection IIB4.

Axioms C1–C5 guarantee the existence of some quantitative probability P agreeing with the comparative probability relation $\gtrsim$. This P is not unique. Any strictly increasing function f of P, $P' = f(P)$, yields a quantitative probability agreeing with $\gtrsim$. Hence we can, if we wish, choose P such that $P(\Omega) = 1$ and $P(F) \geqslant 0$, thereby satisfying the first two Kolmogorov axioms. However, we have said nothing about the concept of comparative probability that would necessitate these choices.

3. An Archimedean Axiom

Modifying a suggestion of Luce [4], we will propose and accept an Archimedean axiom whose purpose is to rule out those $\gtrsim$ which admit infinitely, or infinitesimally, probable events. We first introduce a definition.

Definition. $\{A_i\}$ is an upper scale for A if $A_0 = \varnothing$, $(\forall j \geqslant 0)(\exists B_j, C_j)$ $C_j \gtrsim A$, $B_j \gtrsim A_j$, $C_j \cap B_j = \varnothing$, $A_{j+1} \gtrsim B_j \cup C_j$.

In essence, an upper scale for A is any sequence of sets in which the "gap" between successive terms is of "size" at least A. It is immediate that any subsequence of an upper scale is again an upper scale, and that if $A > B$, then for any upper scale for A there is one at least as long (as many terms) for B. If $\gtrsim$ in the definition of an upper scale is replaced by $\approx$, then we call the resulting scale a standard series; the gaps between successive terms in a standard series are exactly of size A. We can now propose

C6. If $A > \varnothing$, then all upper scales for A are of length finite.

Our version of the Archimedean axiom is strictly stronger than the one proposed by Luce employing standard series in place of upper scales. Consider the following example:

$$\Omega = \{\omega_i\},\ A \in \mathscr{F} \text{ if either } A \text{ or } \bar{A} \text{ is a finite subset of } \Omega,$$

$$(\forall n)\, \omega_n > \bigcup_{j<n} \omega_j \,.$$

It is easily verified that the above-defined $\gtrsim$ satisfies C0–C5 but not C6. For C5 we need only note that $\mathscr{F}$ is a countable collection. For the failure of C6 we observe that $\{A_n = \bigcup_{i \leqslant n} \omega_i\}$ is an infinite-length upper scale for ω_1. However, due to an absence of nontrivial equivalences, all standard series for ω_1 are of length 1.

The independence between C0–C5 and C6 is established in

Theorem 2.

  (a)   C0–C5 $\not\Rightarrow$ C6,
  (b)   C0–C4,   C6 $\not\Rightarrow$ C5.

Upper scales and standard series can have some unintuitive properties. While at first glance it appears reasonable to think of A_n in a standard series for A as n times as probable as A, there can exist standard series $\{A_i\}$ and $\{B_i\}$ for A such that

$$(\exists j)\, A_j \approx B_j\,,\qquad A_{j+1} \approx B_{j+1}\,,\qquad (\exists i) \text{ false } A_i \approx B_i\,.$$

4. An Axiom of Monotone Continuity

This axiom was suggested by Villegas [5] in connection with the existence of countably additive agreeing P. For convenience we denote by $A_i \downarrow A$ the property of a countable set $\{A_i\}$ that

$$(\forall i)\, A_i \supseteq A_{i+1}\,,\qquad \bigcap_{i=1}^{\infty} A_i = A\,.$$

C7. (*Monotone continuity*)　$(\forall i)\ A_i \gtrsim B,\ A_i \downarrow A \Rightarrow A \gtrsim B.$

Unlike C0–C6, we are inclined to view C7 as attractive but not necessary for a characterization of $\gtrsim$. The utility of C7 is apparent from

Theorem 3.

(a)　C0–C6 $\not\Rightarrow$ C7,

(b)　C0–C4, C7 $\Rightarrow$ C5 and C6.

5. Compatibility of CP Relations

Unlike the situation in the usual quantitative probability theories, it is not the case that every pair of CP experiments can be considered as arising from a single underlying CP experiment. To elucidate this we introduce a

Definition. $\mathscr{E}_1 = (\Omega_1, \mathscr{F}_1, \gtrsim_1)$ and $\mathscr{E}_2 = (\Omega_2, \mathscr{F}_2, \gtrsim_2)$ are strongly compatible if there exists $\mathscr{E} = (\Omega, \mathscr{F}, \gtrsim)$ and random variables $x_1, x_2,$

$$x_i : \Omega \to \Omega_i,\qquad (\forall A \in \mathscr{F}_i)\, x_i^{-1}(A) \in \mathscr{F},$$

such that x_i induces $\mathscr{E}_i$; that is,

$$x_i(\Omega) = \Omega_i,\qquad A \gtrsim_i B \Leftrightarrow x_i^{-1}(A) \gtrsim x_i^{-1}(B).$$

As is discussed in Section IIC, there are CP relations that do not admit of any additive agreeing P. It is easily shown under C0–C4 and C7 that no such CP relation is strongly compatible with any CP relation for which there is a countably additive agreeing P whose range contains an interval. A weaker version of compatibility is given in the following.

Definition. $\{\mathscr{E}_\alpha\}$ are mutually weakly compatible if there exist $\mathscr{E}$ and $\{x_\alpha\}$ satisfying: x_α has domain $D_\alpha \in \mathscr{F}$, $x_\alpha(D_\alpha) = \Omega_\alpha$, $\bigcap_\alpha D_\alpha > \varnothing$,

$$(\forall A \in \mathscr{F}_\alpha)\, x_\alpha^{-1}(A) \in \mathscr{F},\qquad A \gtrsim_\alpha B \Leftrightarrow x_\alpha^{-1}(A) \gtrsim x_\alpha^{-1}(B).$$

The requirement of weak compatibility allows for the possibility that some outcomes of, say, $\mathscr{E}_1$ preclude the performance of $\mathscr{E}_2$. This is a very realistic possibility, although one which the quantitative theory had no need to consider. The question of compatible random experiments is a novel one and deserves closer examination than we have given it.

IIC. Compatibility with Finite Additivity

1. Introduction

With the structure thus far exposed for CP we can assert

Theorem 4. If $\succsim$ satisfies C0–C5, then there exists P agreeing with $\succsim$, and there exists a function G of two variables such that

$$A \cap B = \varnothing \Rightarrow P(A \cup B) = G(P(A), P(B)).$$

Furthermore, $G(x, y)$ is symmetric $[G(x, y) = G(y, x)]$, strictly increasing in x, $G(x, P(\phi)) = x$, and associative $[G(G(x, y), z) = G(x, G(y, z))]$.

Acquaintance with the theory of associative functional equations [6] might lead one to expect that by a suitable increasing transformation of P to P' we might find a corresponding G' such that $G'(x, y) = x + y$. That this is not always possible is evident from a simple counterexample due to Kraft *et al.* [7].

Counterexample. Let $\Omega = \{a, b, c, d, e\}$ and $\mathscr{F}$ contain all subsets. Consider the ordering

$$\varnothing \prec a \prec b \prec c \prec ab \prec ac \prec d \prec ad \prec bc \prec e \prec abc \prec bd \prec cd \prec ae \prec abd$$

$$\prec be \prec acd \prec ce \prec bcd \prec abe \prec ace \prec de \prec abcd \prec ade \prec bce \prec abce$$

$$\prec bde \prec cde \prec abde \prec acde \prec bcde \prec abcde = \Omega. \tag{$*$}$$

The relation $\prec$ defined above clearly satisfies C0–C5 and hence admits of a quantitative P. However, there is no such P that is finitely additive. To verify this, assume to the contrary that there exists an additive P assigning probabilities $P(a) = A$, $P(b) = B$, $P(c) = C$, $P(d) = D$, $P(e) = E$. We then see from $(*)$ that

$$A + C < D; \quad A + D < B + C; \quad C + D < A + E.$$

Adding these three inequalities together and subtracting repeated terms yields

$$A + C + D < B + E.$$

Hence $acd \prec be$. However, this is contradicted by $(*)$. Thus there cannot exist an additive quantitative probability agreeing with $\prec$ defined by $(*)$. ∎

2. Necessary and Sufficient Conditions for Compatibility

Necessary and sufficient conditions that ensure that $\gtrsim$ admits of a not necessarily unique, finitely additive representation when Ω is finite have been given first by Kraft *et al.* and then by Scott [8]. Scott's treatment seems preferable and can be more easily extended to infinite Ω. After presenting Scott's theorem we will argue that the hypothesis of the theorem is not an acceptable axiom for CP. It appears that CP relations are not reasonably restricted to only those compatible with additive probability. Nevertheless, it is desirable to have intuitively appealing sufficient conditions for compatibility with additivity, and we present such conditions due to Kraft *et al.*, Luce, and Savage.

As a prelude to Scott's theorem we represent subsets A of Ω by points in a linear vector space $\mathscr{V}$; $\mathscr{V}$ is the linear manifold generated from the set indicator (characteristic) functions

$$I_A(\omega) = \begin{cases} 1 & \text{if} \quad \omega \in A \\ 0 & \text{if} \quad \omega \notin A; \end{cases}$$

that is,

$$\mathscr{V} = \left\{ v : \exists(c_i \in R^1, A_i \subseteq \Omega) \, v = \sum_i c_i I_{A_i} \right\}.$$

The CP relation $\gtrsim$ between subsets of Ω partially defines a relation $\gtrsim'$ between elements of $\mathscr{V}$ as follows: $A \gtrsim B \Rightarrow I_A \gtrsim' I_B$; $c > 0$, $u \gtrsim' v \Rightarrow cu \gtrsim' cv$; $u \gtrsim' v \Rightarrow u + w \gtrsim' v + w$. While we would also wish to assume that $u \gtrsim' 0$, $v \gtrsim' 0 \Rightarrow u + v \gtrsim' 0$, this is not necessarily consistent with $\gtrsim$; to wit, the KPS counterexample. We require an additional hypothesis concerning $\gtrsim$ to guarantee this property for $\gtrsim'$, and one form is provided in the statement of the following.

Theorem (Scott). For finite Ω, the CP relation $\gtrsim$ satisfying C1–C4 is compatible with finite additivity if and only if

$$(\forall u_i, v_i) \left(\sum^n u_i = \sum^n v_i, (\forall i < n)(u_i \lesssim v_i) \Rightarrow u_n \gtrsim v_n \right). \qquad (**)$$

Proof. The proof is based on the convexity of $\{v : v \geqslant' 0\}$ and the hyperplane separation theorem for disjoint convex sets. See Scott [8, pp. 246–247]. ∎

Extension. Scott claims that use of the Hahn–Banach theorem enables him to extend this theorem to the case of infinite Ω.

3. Should CP Be Compatible with Finite Additivity?

When $\succsim$ satisfies C0–C4 a contradiction to (∗∗) can only occur if $\sum u_i$ is not itself an indicator function. In essence, Scott's theorem shows that $\succsim$ is compatible with a finitely additive representation if and only if when $\succsim$ is extended to a collection of objects that are no longer events it does not become possible to find inconsistencies in the ordering of the nonevents. While this observation is of interest, it does not provide a reasonable axiom for a theory of CP concerned solely with events. Why should we be concerned about objects that have no reasonable interpretation in terms of random phenomena? Inconsistencies that arise outside the domain of events, as modeled by sets or indicator functions, do not reflect adversely on the adequacy of the theory when restricted, as it would be, to the domain of events. For example, what in our understanding of CP compels us to eliminate the ordering of (∗) as a CP relation?

While the relationship between additivity and probability is discussed in greater detail in Section IIID, we should note that from the viewpoint of the theory of measurement it is only reasonable to insist upon an additive scale (probability) for uncertainty if this numerical relationship reflects an underlying empirical relationship between uncertainties. Beyond C4, we are unaware of any such relationship between uncertainties that would urge an additive representation. It might be thought that if the interpretation underlying the formal structure of CP is a relative-frequency one, then we will be inexorably led to consider only those CP relations having an additive representation. However, the falsity of this conjecture is demonstrable through the following example of the estimation of a CP relation from relative-frequency data [1].

Let $N_A(n)$ be the number of occurrences of $A \subseteq \Omega$ in n unlinked repetitions of the random experiment $(\Omega, \mathscr{F}, \succsim)$. Conformity between relative-frequency and CP presumably suggests that

$$N_A(n) > N_B(n) \Rightarrow A \succsim B.$$

However, especially for small n, we would not feel obliged to accept

$$N_A(n) = N_B(n) \Rightarrow A \approx B.$$

With this understanding of the estimation of CP from relative-frequency, we could estimate (∗) from the following data:

$$N_a(n)/n = 1/16, \qquad N_b(n)/n = 2/16, \qquad N_c(n)/n = 3/16,$$
$$N_d(n)/n = 4/16, \qquad N_e(n)/n = 6/16.$$

The indications are that the class of reasonable CP relations is significantly larger than the class of relations compatible with a finitely additive representation. However, there seem to be significant difficulties in carrying out the combinatorial analysis necessary to make this statement precise for finite Ω.

4. Sufficient Conditions for Compatibility with Finite Additivity

Sufficient conditions for compatibility of CP with finite additivity have been suggested by Kraft *et al.*, Savage, and Luce. After introducing these conditions we will show that Luce's suggestion is uniformly best, although the other two initially have greater intuitive appeal. Kraft *et al.* suggest use of the property of polarizability, defined as follows.

Definition. A is polarizable if $(\exists A', A'')(A = A' \cup A'', A' \cap A'' = \varnothing, A' \approx A'')$.

The Kraft *et al.* condition is

K. $(\forall A)$ (A is polarizable).

Savage introduced the notions of fine and tight CP relations but we have shown that his weaker assumption of the existence of almost uniform partitions when combined with C5 will suffice for finite additivity.

Definition. An n-fold almost uniform partition $\mathscr{P}_n = \{P_j^{(n)}\}$ is an n-set partition of Ω with the property that the union of no k sets in $\mathscr{P}_n$ is more probable than the union of any $(k + 1)$ sets in $\mathscr{P}_n$.

Savage then hypothesizes

S. For infinitely many n, there exists $\mathscr{P}_n$.

Finally, Luce [4] proposes

L. $(A \cap B = \varnothing, A \gtrsim C, B \gtrsim D) \Rightarrow (\exists C', D', E)(C' \cap D' = \varnothing, E \approx A \cup B, C' \approx C, D' \approx D, E \supseteq C' \cup D')$.

Clearly conditions K and S presume that Ω is infinite, whereas L is compatible with some finite Ω.

It is evident that condition K implies condition S with $\mathscr{P}_n$ existing for all n of the form 2^j. Hence it suffices to treat S and ignore K.

With regard to S, Savage [9] has proven the following.

Theorem (Savage). If $\gtrsim$ satisfies C1–C4 and S, then there exists a unique (to within multiplication by a positive constant) finitely additive function P such that

$$A \gtrsim B \Rightarrow P(A) \geqslant P(B) \qquad \text{(almost agreement)}.$$

Furthermore,

$$(\forall 0 \leqslant \rho \leqslant 1)(\forall A)(\exists B \subseteq A)\,(P(B) = \rho P(A)). \qquad (***)$$

To close the gap between almost agreement (implication in one direction) and agreement (implication in both directions) we introduce

Theorem 5. If a finitely additive P satisfies $(***)$ and almost agrees with $\gtrsim$ satisfying C0–C4, then P agrees with $\gtrsim$ if and only if $\gtrsim$ satisfies C5.

Remark. In this case the gap between almost and strict agreement is solely due to the possible nonexistence of any (not necessarily additive) agreeing function.

Corollary. If $\gtrsim$ satisfies C0–C5 and S, then there exists a unique (to within multiplication by a positive constant) finitely additive P satisfying $(***)$ and agreeing with $\gtrsim$.

This corollary informs us as to the most that can be expected of those CP relations compatible with the hypothesis of infinitely many almost uniform partitions.

With respect to condition L, Luce [4] has shown the following.

Theorem (Luce). If the CP relation $\gtrsim$ satisfies C0–C4, C6, and L, then there exists a unique (to within multiplication by a positive constant) finitely additive function P agreeing with $\gtrsim$.

To demonstrate that (C0–C4, C6, L) provide a uniformly better characterization of finite additivity than does (C0–C5, S), we establish

Theorem 6. (C0–C5, S) $\Rightarrow$ (C0–C4, C6, L), but the converse implication is not valid.

To date, the set (C0–C4, C6, L) seems to be the best published characterization of the subset of CP relations compatible with finite additivity, where by "best" we refer to the intuitive acceptability of the axioms as well as to the size of the subset. Some insight into C6, L is provided by the discussion in Section IIID of the empirical meaning of additivity for a probability measurement scale. That discussion leads us to conclude that (C1–C4, C6, L) is too stringent a characterization of those CP relations for which additive P is empirically meaningful.

Results on the related problem of the existence of almost agreeing finitely additive P are available in Fishburn [10] and will not concern us here.

IID. Compatibility with Countable Additivity

In view of the widespread acceptance of the Kolmogorov axioms (see Section IIIA) for probability, it is of interest to characterize the subclass of CP relations that are compatible with countably additive agreeing P. As is well known in probability theory, countable additivity is equivalent to finite additivity and the following continuity condition:

$$B_i \downarrow \varnothing \Rightarrow \lim_{i \to \infty} P(B_i) = 0.$$

The continuity condition suggests the following weakened version of C7 as an axiom for CP.

C8. $(\forall \{B_i\}) \left(B_i \downarrow \varnothing \Rightarrow \bigcap_{i=1}^{\infty} \{A : \varnothing < A \leqslant B_i\} = \varnothing \right).$

The role of C8 can best be seen from

Theorem 7. If $\gtrsim$ admits of an agreeing finitely additive P, then P is countable additive if and only if $\gtrsim$ satisfies C8.

Corollary. If $\gtrsim$ satisfies C1–C4, C7, and L, then there exists a unique agreeing P satisfying the Kolmogorov axioms.

Villegas [5] has proposed a characterization of a subset of CP relations compatible with countable additivity. After introducing the hypothesis that there are no atoms,

$$(\text{atomless}) \quad (\forall A > \varnothing)(\exists B \subset A)(A > B > \varnothing),$$

he then establishes

Theorem (Villegas). If $\gtrsim$ satisfies C0–C4 and C7 and is atomless, then there is a unique agreeing P satisfying the Kolmogorov axioms.

Unfortunately Villegas' simple conditions yield a theory of countable additivity that is strictly less powerful than that of the preceding corollary. It follows from Villegas' Theorem 5 that his axioms imply C0–C7, L, whereas the converse is false.

Our discussion of the formal structure of CP is incomplete in at least two respects.

(1)   Do axioms C0–C7 suffice to characterize CP relations?
(2)   What are suitable definitions for the associated concepts needed to produce a rich theory of CP?

Concerning (1) we are doubtful, but have nothing to add. It is to consideration of (2) that we now turn.

IIE. Comparative Conditional Probability

1. Comparative Conditional Probability as a Ternary Relation

Our first attempt at comparative conditional probability (CCP) views it as a ternary relation $\succsim^*$ (TCCP) on $\mathscr{F} \times \mathscr{F} \times \mathscr{G}$ where $\mathscr{F}$ is a field of events and $\mathscr{G} \subset \mathscr{F}$; a typical statement is written "$A/B \succsim^* C/B$" and read "A given B is at least as probable as C given B," for A, $C \in \mathscr{F}$ and $B \in \mathscr{G}$. Given any $B \in \mathscr{G}$ we define $\succsim_B$ through

$$(\forall A, C \in \mathscr{F})\, A \succsim_B C \Leftrightarrow A/B \succsim^* C/B.$$

We characterize TCCP through two axioms.

TCC1. $(\forall B \in \mathscr{G})\, \succsim_B$ satisfies C0–C6.

TCC2. $(\forall B, C \in \mathscr{G})\, B \supseteq C \Rightarrow (A/C \succsim^* D/C \Leftrightarrow A \cap C/B \succsim^* D \cap C/B)$

A simple consequence of TCC2 is that if $\bigcup_{G \in \mathscr{G}} G = H \in \mathscr{G}$, then $\succsim^*$ is determined by $\succsim_H$ through

$$(\forall A, C \in \mathscr{F})(\forall B \in \mathscr{G})\, A/B \succsim^* C/B \Leftrightarrow A \cap B \succsim_H C \cap B.$$

While TCCP is a weaker notion than the quaternary relation to be discussed next, it nevertheless is adequate for some forms of inference. Given an observation $B \in \mathscr{G}$ we are enabled to determine the posterior ordering of elements of $\mathscr{F}$ and, for example, to identify the *a posteriori* most probable of a collection of events corresponding to hypotheses. Furthermore, while there is always a TCCP relation $\succsim^*$ corresponding to any CP relation $\succsim$, the same does not hold true for the more highly structured quaternary relation.

2. Comparative Conditional Probability as a Quaternary Relation

Early investigations of quaternary comparative conditional probability (QCCP) were initiated by Keynes [11] and Koopman [12]. As these investigations are closely related to the concept of logical probability, we defer their examination to Chapter VII. Recent research on QCCP and its relation to quantitative conditional probability has been reported by Domotor [13] and Luce [14]. Our approach was motivated by the earlier work of Renyi [15] (see Section IIIF) on conditional probability and leads to a set of axioms different from those of Domotor and Luce. In this section we restrict ourselves to consideration of the formal structure of the QCCP relation denoted "$A/B \succsim^* C/D$" and read "A given B is as least as probable as C given D." In "A/B" we assume that A and B are subsets of Ω; A is in a field $\mathscr{F}$; and B is in a subset $\mathscr{G}$ of $\mathscr{F}$ that is closed under intersection and does not contain events equivalent to $\varnothing$.

The first seven axioms for QCCP are direct counterparts of C0–C6.

QCC0. (*Nontriviality*) $\quad \Omega/A \succ^* \varnothing/A.$

QCC1. (*Comparability*) $\quad A/B \succsim^* C/D \quad$ or $\quad C/D \succsim^* A/B.$

QCC2. (*Transitivity*) $\quad A/B \succsim^* C/D, \quad C/D \succsim^* E/F \Rightarrow A/B \succsim^* E/F.$

QCC3. (*Improbability of impossibility*) $\quad A/B \succsim^* \varnothing/D.$

QCC4. (*Disjoint unions*) $\quad$ If $A \cap B = C \cap D = \varnothing$ then $A/E \succsim^* C/F$

$$B/E \succsim^* D/F \Rightarrow A \cup B/E \succsim^* C \cup D/F,$$

and if in addition

$$A/E \succ^* C/F \quad \text{then} \quad A \cup B/E \succ^* C \cup D/F.$$

QCC5. (*Compatibility with quantitative representations*) If $\mathscr{H}$ is the collection of pairs $\{A/B : A \in \mathscr{F}, B \in \mathscr{G}\}$, $\mathscr{T}$ the order topology induced by $\succ^*$, then $(\mathscr{H}, \mathscr{T})$ has a countable base.

Rewriting the definition of $\{A_i/B\}$ as an upper scale for A/B we have

$$A_0/B = \varnothing/B$$

$$(\forall j \geqslant 0)(\exists B_j, C_j) \; C_j/B \succsim A/B, \quad B_j/B \succsim A_j/B,$$

$$C_j \cap B_j \cap B = \varnothing, \quad A_{j+1}/B \succsim B_j \cup C_j/B.$$

We then postulate

QCC6. (*Archimedean order*) If $A/B >^* \varnothing/B$, then any upper scale for A/B has only finitely many terms.

The final, key axiom relates to the "product rule" and is central to inference.

QCC7.

(a) $A_1/A_2 \cap A_3 \gtrsim^* B_1/B_2 \cap B_3 , A_2/A_3 \gtrsim^* B_2/B_3 \Rightarrow A_1 \cap A_2/A_3$
$\gtrsim^* B_1 \cap B_2/B_3;$

(b) $A_1/A_2 \cap A_3 \gtrsim^* B_2/B_3 , A_2/A_3 \gtrsim^* B_1/B_2 \cap B_3 \Rightarrow A_1 \cap A_2/A_3$
$\gtrsim^* B_1 \cap B_2/B_3;$

(c) $A_2/A_3 \gtrsim^* B_2/B_3 , \quad A_2/A_3 >^* \varnothing/B, \quad A_1/A_2 \cap A_3$
$>^* B_1/B_2 \cap B_3 \Rightarrow A_1 \cap A_2/A_3 >^* B_1 \cap B_2/B_3 .$

An understanding of the role of QCC7 is available from

Theorem 8. If $\gtrsim^*$ satisfies QCC1, QCC2, QCC5, then there exists some agreeing function P. There exists a function G of two variables such that

(1) $P((A \cap B) \mid C) = G(P(B/C), P(A/B \cap C)),$
(2) $G(x, y) = G(y, x)$ (symmetry),
(3) $G(x, y)$ is increasing in x for $y > P(\varnothing/B),$
(4) $G(x, G(y, z)) = G(G(x, y), z)$ (associativity),
(5) $G(P(\Omega/B), y) = y,$
(6) $G(P(\varnothing/B), y) = P(\varnothing/B),$

if and only if $\gtrsim^*$ also satisfies QCC7.

Corollary. If the QCCP relation $\gtrsim^*$ satisfies QCC1–QCC7, then there exists $P : \{A/B\} \to [0, 1]$, there exists G, $P(\varnothing/A) = 0$, $P(A/A) = 1$, $P(A \cap B/C) = G(P(B/C), P(A/B \cap C))$, $P(A/B) = P(A \cap B/B)$, and for fixed B, $P(\cdot/B) : \mathscr{F} \to [0, 1]$ satisfies C0–C6.

3. Relationship between QCCP and CP

We are now able to define a binary relation of CP from the quaternary relation of QCCP. Select a fixed event $B_0 \in \mathscr{G}$, the usual choice being Ω. We then define $A \gtrsim_{B_0} B \Leftrightarrow A/B_0 \gtrsim^* B/B_0$. It is straightforward to show that $\gtrsim_{B_0}$ as so defined has the properties C0–C6.

It is not generally possible to define a QCCP structure in terms of CP, as is commonly done in quantitative probability. To define $\gtrsim^*$ from $\gtrsim$ we would presumably start by defining

$$A/C \gtrsim^* B/D \quad \text{if} \quad D \gtrsim C \quad \text{and} \quad A \cap C \gtrsim B \cap D.$$

We must then extend the ordering $\gtrsim^*$ to other pairs. This extension need not be possible, and when possible it need not follow uniquely from $\gtrsim$. Kaplan [1] has constructed an example of a CP order $\gtrsim$ on a sample space Ω of 10 atoms that has the property

$$(\forall \gtrsim^* \text{ satisfying QCC0–QCC4 } (\exists A, B)\ A \succ B,\ B/\Omega \gtrsim^* A/\Omega).$$

Hence there are CP relations that admit of no corresponding QCCP relation, and this incompatibility is fundamental to QCCP as it is based only on QCC0–QCC4. A sufficient condition for the existence of a corresponding QCCP relation is that the CP relation $\gtrsim$ be extendible to the case of two independent repetitions, $\gtrsim_2$ on $\Omega \times \Omega$ (see Section IIF). In this eventuality we merely define

$$A/B \gtrsim^* C/D \Leftrightarrow (A \cap B, D) \gtrsim_2 (C \cap D, B).$$

Specific counterexamples to the unicity of $\gtrsim^*$ corresponding to $\gtrsim$ are easily constructed in the class of CP relations admitting nonunique agreeing probability measures.

The negative result concerning the compatibility of CP and QCCP suggests that not all of the quantitative formulations of inference problems are possible in the framework of CP. Formulations in which the problem resolution depends on QCC1 are the most likely ones to have to be modified. Furthermore this result reflects adversely on the oft-espoused position that conditional probability is more fundamental than unconditional (absolute) probability.

4. Bayes Theorem

It is possible to derive a weak restatement of Bayes theorem for QCCP.

Theorem 9. If $\gtrsim^*$ satisfies QCC1–QCC7, $\Omega \in \mathscr{G}$, then there exist P agreeing with $\gtrsim^*$ and real-valued functions F and G of two variables such that if $A/\Omega \succ \varnothing/\Omega$, then

$$P(B/A) = F(G(P(A/B), P(B/\Omega)), P(A/\Omega)).$$

Furthermore, G has properties (1)–(6) listed in Theorem 8, and $F(x, y)$ is increasing in x and decreasing in y.

Theorem 9 is the comparative analog of Bayes theorem; it informs us that the probability of B/A is determinable from the probabilities of A/B, A/Ω, and B/Ω.

5. Compatibility of CCP with Kolmogorov's Quantitative Probability

As we would expect from Sections IIC and IID, we cannot assert that if $\succsim^*$ satisfies QCC1–QCC7, then there exists some agreeing P of two set-valued variables which is a measure in one variable and for which it is simultaneously true that

$$P(A \cap B/C) = P(A/B \cap C)\,P(B/C).$$

It is possible to strengthen the axioms we have presented so that $\succsim^*$ will be compatible with Kolmogorov-type conditional probabilities. Examples of strengthened systems are provided by Domotor [13] and Luce [14]. However, for reasons presented in Sections IIC and IIID we find that such axioms are unnecessarily restrictive of our informal understanding of QCCP.

IIF. Independence

1. Independent Events

The notion of independence is of great significance for both the practice and theory of probability. As Kolmogorov[†] [16] has said, "We thus see, in the concept of independence, at least the germ of the peculiar type of problem in probability theory." Without entering explicitly into interpretive questions concerning the concept of independence, we wish to consider a formal structure for the informal notion that the occurrence of A is unrelated to, unlinked with, or uninfluenced by the occurrence of B. This formal structure is to relate independence to CP as the scale for the occurrence of events. Formal independence will be taken to be a binary relation between events denoted by "$A \perp\!\!\!\perp B$" and read as "A is independent of B." The axioms

---

[†] A. Kolmogorov, *Foundations of the Theory of Probability*, p. 9. Bronx, New York: Chelsea, 1956.

we present concern the use of the symbol $\perp\!\!\!\perp$, the relation of $\perp\!\!\!\perp$ to set operations, and the relation of $\perp\!\!\!\perp$ to CP. Unlike the approach originally taken by Kolmogorov, we are attempting to axiomatize independence between events and not independence between experiments. Some remarks concerning the adoption of independent experiments as the fundamental concept are made at the conclusion of this section. While it appears to be easier to formalize independent experiments than it is to formalize independent events, the difficulties encountered are instructive.

The first axiom is

I1. (*Symmetry*) $A \perp\!\!\!\perp B \Rightarrow B \perp\!\!\!\perp A.$

If A is unrelated to B, then B is unrelated to A. Although the usual definition of stochastic independence accepts I1, an examination of independence from the computational-complexity viewpoint presented in Section VH casts a little doubt on its acceptability.

The next three axioms relate $\perp\!\!\!\perp$ to sets and set operations.

I2. $A \perp\!\!\!\perp \Omega.$

I3. $A \perp\!\!\!\perp B \Rightarrow A \perp\!\!\!\perp \bar{B}.$

I4. $B \cap C = \varnothing,\ \ A \perp\!\!\!\perp B,\ \ A \perp\!\!\!\perp C \Rightarrow A \perp\!\!\!\perp (B \cup C).$

Axiom I2 expresses the informal belief that the occurrence of any event is unrelated to the (necessary) occurrence of the certain event. Axiom I3 asserts the reasonable understanding that if the occurrences of A and B are unrelated, then so must be their nonoccurrences. Axiom I4, in effect, asserts that if A is unrelated to both B and C, and B and C are mutually exclusive, then A is unrelated to any Boolean function of B and C. It is criticized in Section VH.

A relation between $\perp\!\!\!\perp$ and the CP relation $\gtrsim$ is given by

I5. $A \perp\!\!\!\perp B,\ \ C \perp\!\!\!\perp D,\ \ B \gtrsim D \Rightarrow (A \gtrsim C \Rightarrow A \cap B \gtrsim C \cap D),$

$(A \succ C, B \succ \varnothing \Rightarrow A \cap B \succ C \cap D).$

I5 is the key axiom relating $\gtrsim$ and $\perp\!\!\!\perp$.

With the structure thus far exposed for $\perp\!\!\!\perp$ we can state

Theorem 10. If $\gtrsim$ satisfies C1–C5, and $\perp\!\!\!\perp$ satisfies I1 and I5, then there exists P agreeing with $\gtrsim$ and G such that

$$A \perp\!\!\!\perp B \Rightarrow P(A \cap B) = G(P(A), P(B)).$$

Furthermore, $G(x, y)$ is symmetric and increasing in x for $y > P(\varnothing)$.

Other consequences of I1–I5 include the following expected results:

(a) $A \perp\!\!\!\perp B$, $A \supseteq B \Rightarrow A \approx \Omega$ or $B \approx \varnothing$.

(b) $A \cap B = \varnothing$, $A \perp\!\!\!\perp B \Rightarrow A \approx \varnothing$ or $B \approx \varnothing$.

(c) $A \perp\!\!\!\perp B$, $A \perp\!\!\!\perp C$, $A \perp\!\!\!\perp (B \cap C) \Rightarrow A \perp\!\!\!\perp (B \cup C)$.

It does not follow from I1–I5 that G as defined in Theorem 10 is necessarily associative. Of course, the Kolmogorov definition of independence does assure us of the associativity of G. To guarantee associativity we suggest axiom

I6. $A \perp\!\!\!\perp B \cap C$, $B \perp\!\!\!\perp C$, $A' \cap B' \perp\!\!\!\perp C'$, $A' \perp\!\!\!\perp B'$, $A \approx A'$,

$\quad B \approx B'$, $C \approx C' \Rightarrow A \cap B \cap C \approx A' \cap B' \cap C'$.

This axiom does not seem to us to be particularly compelling within the context of a theory of pairwise independent events. Its role, however, is apparent from the following.

Lemma 1. G as defined in Theorem 10 is associative, when all terms are defined, if and only if $\perp\!\!\!\perp$ satisfies I6.

Another possible property of $\perp\!\!\!\perp$ is

I7. $A \perp\!\!\!\perp B$, $B \perp\!\!\!\perp C$, $A \perp\!\!\!\perp B \cap C \Rightarrow C \perp\!\!\!\perp A \cap B$.

I7 is a theorem in the Kolmogorov system, but is independent of I1–I6.

A significant discrepancy between our notion of (comparative) independence and the Kolmogorov notion of (stochastic) independence is that $\perp\!\!\!\perp$ is not determined by $\gtrsim$. Our axioms I1–I7 merely assert compatibility requirements between independence and comparative probability. From

$$A \perp\!\!\!\perp B \Rightarrow P(A \cap B) = G(P(A), P(B)),$$

we cannot conclude that

$$P(A \cap B) = G(P(A), P(B)) \Rightarrow A \perp\!\!\!\perp B.$$

The locations of A, B, and $A \cap B$ in the $\gtrsim$ order chain do not determine the independence of A and B although they can preclude it. Restated in more familiar terms it is as if product factorization of probability is necessary but not sufficient for independence. Interestingly, a common defense of the Kolmogorov definition of independence through the relative-frequency interpretation (see Sections IIIG and VH) only establishes necessity and not sufficiency.

An analogy that may make these observations more palatable is to compare independence with disjointness. In the usual probability theory

$$A \cap B = \varnothing \Rightarrow P(A \cup B) = P(A) + P(B).$$

However, if for some C and D it happened that $P(C \cup D) = P(C) + P(D)$ we would not conclude that C and D were disjoint.

The practical import of the absence of sufficiency may be small. Generally we work from assumptions of independence to probability models. It is less common to start with a probability model and conclude with statements of independence. Nevertheless, this happens and is mentioned in Section VH.

A step in the direction of relating $\gtrsim$ to $\perp\!\!\!\perp$ is given by

18. $A \approx C,\ \ B \approx D,\ \ A \cap B \approx C \cap D,\ \ A \perp\!\!\!\perp B \Rightarrow C \perp\!\!\!\perp D.$

Axiom I8 asserts that if for some A and B, $P(A) = x$, $P(B) = y$, $P(A \cap B) = z$, $A \perp\!\!\!\perp B$, then any other pair of events C, D for which $z = G(x, y)$ must also be independent. Hence, under I8 $\gtrsim$ can occasionally establish the independence of events. We do not find strong grounds upon which to base an acceptance of I8.

A final axiom relating independence and limits is

19. $A_n \uparrow A,\ \ B_n \uparrow B,\ \ A_n \perp\!\!\!\perp B_n \Rightarrow A \perp\!\!\!\perp B.$

Since we can arbitrarily well approximate A by some A_n and B by some B_n, and A_n is unrelated to B_n, it seems reasonable to assert that A is unrelated to B. Of course, an axiom concerning limits cannot be "justified" by appeal to our finitistic, experientially based intuition concerning unrelatedness or unlinkedness.

It is of course patent from the preceding discussion that our characterization of $\perp\!\!\!\perp$ through I1–I9 is too weak to ensure that $\perp\!\!\!\perp$ is compatible with the usual Kolmogorov definition of independence. Domotor [13, pp. 59–73] presents a much stronger set of axioms for $\perp\!\!\!\perp$ that do lead to the usual definition. However, to achieve this end he works with a quaternary relational structure that is not that of CP and introduces several axioms whose justification is not apparent. Nevertheless his work contains many interesting results on independence.

2. Mutually Independent Events

We can easily extend our treatment of pairwise independence to cover the n-ary relation $\perp\!\!\!\perp_n(A_1, ..., A_n)$ that $A_1, ..., A_n$ are mutually independent. For $n = 2$ we identify $\perp\!\!\!\perp_2(A_1, A_2)$ with $A_1 \perp\!\!\!\perp A_2$. For $n > 2$, we recursively define $\perp\!\!\!\perp_n$ as follows.

Definition. $\perp\!\!\!\perp_n(A_1, ..., A_n)$ if $(\forall j)$ $\perp\!\!\!\perp_{n-1}(A_1, ..., A_{j-1}, A_{j+1}, ..., A_n)$ and $A_j \perp\!\!\!\perp_2 (\bigcup_{i \neq j} A_i)$.

It is easily seen from the definition of $\perp\!\!\!\perp_n$ that as expected, all subsets of $\{A_1, ..., A_n\}$ are also mutually independent collections.

3. Independent Experiments

As indicated at the outset of our discussion of independence we treated independence as a binary relation between events. Kolmogorov and others have preferred to start with independence as an n-ary relation between experiments, where experiments are represented as a field of events. The advantage of our approach is that it deals with the independence relation between the much simpler notion of a pair of events. Yet, it might be argued that despite the relative complexity of a collection of experiments, this is intuitively the proper domain for independence. In practice we may more commonly identify experiments as being independent (e.g., the outputs from two gassy tubes in separate circuits, or two roulette wheels on separate tables) than identify events as independent.

The formal equivalence between the notions $\perp\!\!\!\perp^*$ of mutually independent experiments $\mathscr{E}_1, ..., \mathscr{E}_n$, and $\perp\!\!\!\perp_n$ of mutually independent events $A_1, ..., A_n$ can be delineated as follows:

$$\perp\!\!\!\perp_n^*(\mathscr{E}_1, ..., \mathscr{E}_n) \quad \text{if} \quad (\forall i)(\forall A_i \in \mathscr{E}_i)\left(\perp\!\!\!\perp_n(A_1, ..., A_n)\right)$$

or

$$\perp\!\!\!\perp_n(A_1, ..., A_n) \quad \text{if} \quad (\forall i)\left(\mathscr{E}_i = \{\varnothing, A_i, \bar{A}_i, \Omega\}\right)\left(\perp\!\!\!\perp_n^*(\mathscr{E}_1, ..., \mathscr{E}_n)\right).$$

In view of these correspondences we do not directly axiomatize $\perp\!\!\!\perp^*$. Additional results concerning the formal structure of independence can be found in the works of Domotor [13, pp. 59–73] and Kaplan [17].

4. Concluding Remarks

Our formal development of independence has its roots solely in introspection concerning the meaning of unlinkedness between events whose outcomes are uncertain. The concept we axiomatized was of an epistemological and uninterpreted character. In Section IIH we present an axiom (E5) relating expectations and randomizations that could, after some adaptation, be reversed to yield a pragmatic, decision-oriented definition of independence. In effect we would consider experiments $\mathscr{E}_1$ and $\mathscr{E}_2$ to be independent if our preferences for gambles or acts based on the outcome of one experiment are unchanged when we are informed of the outcome of the other experiment.

In the more traditional epistemological vein, Section VH contains a computational-complexity-based characterization of what it means for the outcome of one experiment to be uninformative about the outcome of another experiment leads to a definition of independence that only approximately satisfies I1–I8. Yet this new definition clearly points up the inadequacy of any relative-frequency-based definition of independence and the lack of sufficiency in the usual probabilistic definition of independence.

IIG. Application to Decision-Making

1. Formulation of the Decision Problem

Notwithstanding the relative weakness of CP statements, it is possible to pose and resolve interesting decision problems in a CP framework. The approach to decision making we tentatively propose is intended more as an illustration of our assertion than as a thoroughly considered "best" formulation of decision-making in CP. Much more research is needed before we can claim a satisfactory complete analysis of decision-making in CP.

Our initial model of a CP decision problem rests upon the following definitions, where to avoid needless technicalities, all sets will be assumed finite unless otherwise indicated.

$\mathscr{S} = \{s\}$ set of states of nature, hypotheses, parameter values.

$\gtrsim$ CP relation between subsets of $\mathscr{S}$ satisfying C0–C6.

$\mathscr{P}$ infinite set of (abstract) payoffs.

$\gtrsim_{\mathscr{P}}$ preference relation between elements of $\mathscr{P}$.

$\alpha : \mathscr{S} \to \mathscr{P}$ an act.

$\{ p_i \mid A_i \}$ the act assigning payoff $p_i \in \mathscr{P}$ to the element $A_i \subseteq \mathscr{S}$ of a partition of $\mathscr{S}$.

$\mathscr{A} = \{\alpha\}$ set of all acts.

$\gtrsim_{\mathscr{A}}$ preference relation between acts.

Hence the consequence of an act $\alpha \in \mathscr{A}$ when $s \in \mathscr{S}$ is the true state is $\alpha(s) \in \mathscr{P}$, and our prior information concerning which state is true is contained in $\succsim$. The decision problem is to determine an optimal act or, more generally, to determine which of any pair of acts is the preferable.

If we wish to account explicitly for the role of observations in decision-making, then we can augment the preceding problem set up with the following terms.

$\Omega = \{\omega\}$ set of observations.

$\{\succsim_\omega , \omega \in \Omega\}$ indexed set of CP relations between subsets of $\mathcal{S}$ that individually satisfy C0–C6.

$d : \Omega \to \mathcal{O}$ decision function or rule.

$\mathcal{D} = \{d\}$ set of all decision functions.

$\succsim_{\mathcal{D}}$ preference relation between elements of $\mathcal{D}$.

Having observed $\omega \in \Omega$ we know that the CP relation between subsets of $\mathcal{S}$ is given by $\succsim_\omega$, and using decision function $d \in \mathcal{D}$ we take act $d(\omega) \in \mathcal{O}$. The decision problem now is to identify an optimal decision function or, more generally, to determine which of any pair of decision functions is the preferable. In fact we will only achieve a partial ordering $\succsim_{\mathcal{D}}$ enabling us to determine admissibility of a decision function.

Other formulations of the decision problem in CP are possible. For example, we may assume a CP relation $\succsim$ on $\Omega \times \mathcal{S}$ instead of the weaker assumption $\{\succsim_\omega , \omega \in \Omega\}$, where

$$(\forall \omega \in \Omega)(\forall A, B \subseteq \mathcal{S}) \, (\omega, A) \succsim (\omega, B) \Leftrightarrow A \succsim_\omega B.$$

The disadvantage of this reformulation is that it presumes additional prior knowledge. The advantage of this formulation, as noted in Fine [18], is that unlike our present development we are enabled to compare every pair of decision functions.

2. Axioms for Rational Decision-Making

The approach that we take in choosing between acts is to embed $(\mathcal{O}, \succsim_{\mathcal{O}})$ in $(\mathcal{P}, \succsim_{\mathcal{P}})$. The homomorphism between $(\mathcal{O}, \succsim_{\mathcal{O}})$ and $(\mathcal{P}, \succsim_{\mathcal{P}})$ is characterized through axioms of rationality and axioms of structure or scaling. We then extend the resolution of the decision problem between acts to a partial resolution of the decision problem between rules through a rationality axiom of dominance. This enables us to partially order the decision functions or rules and to isolate the admissible rules. Our solutions will be formal and uninterpreted in that we will not account for the decision maker's ability to satisfy the structural axioms. If, for example, the decision-maker is human and is able, through introspection, to scale subjectively all acts in terms of payoffs, then the overall theory will be one of subjective decision-making with CP that can be compared with the development of Chapter VIII.

We present four rationality axioms and two structural axioms that lead to a representation for $(\mathcal{A}, \succsim_a)$ in $(\mathcal{P}, \succsim_{\mathcal{P}})$. Our first axiom is an obvious requirement for a preference.

DCP1. $\succsim_a$ is transitive and reflexive.

The next two rationality axioms imply that we prefer acts yielding better payoffs on more probable events.

DCP2. $\bar{p} \succ_{\mathcal{P}} p$, $\quad A \succsim B \Leftrightarrow \bar{p} \succ_{\mathcal{P}} p$, $\quad$, $\quad \{\bar{p} \mid A, p \mid \bar{A}\} \succsim_a \{\bar{p} \mid B, p \mid \bar{B}\}$

DCP3. $A \succ \varnothing$, $\quad p \succsim_{\mathcal{P}} q \Leftrightarrow A \succ \varnothing$, $\quad \{p \mid \mathcal{S}\} \succsim_a \{q \mid A, p \mid \bar{A}\}$.

The fourth rationality axiom is a version of Savage's sure-thing principle. It implies that if two acts are identical over $A \subset \mathcal{S}$, then the preference between them is solely determined by their behavior on $\bar{A}$.

DCP4. $(\forall s \in A)\,(\alpha_1(s) = \alpha_2(s)) \Rightarrow (\forall \alpha_3 \in \mathcal{A})\, \alpha_1 \succsim_a \alpha_2$

$$\Leftrightarrow \{\alpha_3 \mid A, \alpha_1 \mid \bar{A}\} \succsim_a \{\alpha_3 \mid A, \alpha_2 \mid \bar{A}\}.$$

While DCP1–DCP4 partially characterize rationality by requiring a certain consistency between preferences for acts, the next two axioms are structural in that they require the decision maker to have sufficiently many preferences.

DCP5. $\succsim_{\mathcal{P}}$ is a complete order on $\mathcal{P}$.

DCP6. $(\forall p_1, p_2 \in \mathcal{P})(\forall A \supset B)(\exists p_3 \in \mathcal{P})\, \{p_1 \mid B, p_2 \mid \bar{B}\} \sim_a \{p_3 \mid A, p_2 \mid \bar{A}\}$.

While DCP5 may be acceptable when, say, $\mathcal{P}$ is money, it may be unacceptable when $\mathcal{P}$ involves a great variety of disparate objects. DCP6 is a scaling axiom known as payoff solvability; it implies that there is always a compensating price for any change in uncertainty.

Turning to the problem of choosing between decision functions, we partially relate $(\mathcal{D}, \succsim_{\mathcal{D}})$ and $(\mathcal{A}, \succsim_a)$ through an axiom of dominance. Let $\succsim_a^{\omega}$ denote the ordering of acts when $\succsim_{\omega}$ is the CP ordering of subsets of $\mathcal{S}$ corresponding to an observation ω.

DCP7.

(a) $(\forall \omega \in \Omega)\, d_1(\omega) \gtrsim_{\alpha}^{\omega} d_2(\omega) \Rightarrow d_1 \gtrsim_{\mathscr{D}} d_2;$

(b) $(\forall \omega \in \Omega)\, d_1(\omega) \gtrsim_{\alpha}^{\omega} d_2(\omega),\quad (\exists \omega)\, d_1(\omega) \succ_{\alpha}^{\omega} d_2(\omega) \Rightarrow d_1 \succ_{\mathscr{D}} d_2 .$

If d_1 always yields acts that are preferable to those of d_2, then we prefer d_1 to d_2. While reasonable, DCP7 does rule out factors such as cost of implementation in determining the preferability of rules.

3. Representations of $\gtrsim_\alpha$ and $\gtrsim$

A basic implication of DCP1–DCP4 is given by

Lemma 2.   If DCP1, DCP3, DCP4, $\{E_i\}$ is any finite partition of $\mathscr{S}$, and

$$\alpha = \{p_i \mid E_i\}, \qquad \alpha' = \{p_i' \mid E_i\},$$

then

(a) $(\forall i)\, p_i \gtrsim_{\mathscr{P}} p_i' \Rightarrow \alpha \gtrsim_\alpha \alpha',$

(b) $(\forall i)(p_i \gtrsim_{\mathscr{P}} p_i'), (\exists j)\, (p_j \succ_{\mathscr{P}} p_j', E_j \succ \varnothing) \Rightarrow \alpha \succ_\alpha \alpha'.$

The key result embedding $(\mathscr{O\!l}, \gtrsim_\alpha)$ in $(\mathscr{P}, \gtrsim_{\mathscr{P}})$ is

Theorem 11.   If DCP1, DCP3, DCP4, DCP6, then

$$(\exists \pi : \mathscr{O\!l} \to \mathscr{P})\, \alpha \gtrsim_\alpha \alpha' \Leftrightarrow \pi(\alpha) \gtrsim_{\mathscr{P}} \pi(\alpha').$$

In effect, any act α is equivalent to some constant act $\{\pi(\alpha) \mid \mathscr{S}\}$, with the preferences between acts α and α' being identical to the preferences between the payoffs $\pi(\alpha)$ and $\pi(\alpha')$.

The implications of DCP1–DCP7 for a partial ordering between decision functions are indicated in

Theorem 12.   If each $\gtrsim_\alpha^\omega \in \{\gtrsim_\alpha^\omega,\ \omega \in \Omega\}$ satisfies DCP1, DCP3, DCP4, DCP6, and if DCP7 holds, then

(a) $(\forall \omega \in \Omega)(\exists \pi_\omega : \mathscr{O\!l} \to \mathscr{P})\, \alpha \gtrsim_\alpha^\omega \alpha' \Leftrightarrow \pi_\omega(\alpha) \gtrsim_{\mathscr{P}} \pi_\omega(\alpha').$

(b) $(\forall \omega \in \Omega)\, \pi_\omega(d_1(\omega)) \gtrsim_{\mathscr{P}} \pi_\omega(d_2(\omega)) \Rightarrow d_1 \gtrsim_{\mathscr{D}} d_2 .$

(c) $(\forall \omega \in \Omega)\, \big(\pi_\omega(d_1(\omega)) \gtrsim_{\mathscr{P}} \pi_\omega(d_2(\omega))\big), (\exists \omega)\big(\pi_\omega(d_1(\omega)) \succ_{\mathscr{P}} \pi_\omega(d_2(\omega))\big)$
$\Rightarrow d_1 \succ_{\mathscr{D}} d_2 .$

Definition [19]. d_2 is admissible in $\mathscr{D}' \subseteq \mathscr{D}$ if $d_2 \in \mathscr{D}'$ and there is no $d_1 \in \mathscr{D}'$ such that Theorem 12(c) holds.

Definition. d' is optimal in $\mathscr{D}' \subseteq \mathscr{D}$ if $d' \in \mathscr{D}'$ and

$$(\forall d \in \mathscr{D}') \, d' \gtrsim_{\mathscr{D}} d.$$

Hence, Theorem 12 enables us to identify the admissible rules in some $\mathscr{D}'$ but not necessarily those that are also optimal in $\mathscr{D}'$.

Knowing only the admissible rules is tantamount to knowing the optimal rules if and only if all of the admissible rules are known to be equivalent. In our formulation this requires that for any two admissible rules d_1 and d_2,

$$(\forall \omega) \, \pi_\omega(d_1(\omega)) \approx_{\mathscr{P}} \pi_\omega(d_2(\omega)).$$

An example of this possibility is provided by decision-making with the so-called ideal observer loss function:

$$(\exists \bar{p} >_{\mathscr{P}} p) \, \mathcal{O}\!\mathcal{U}' = \{\alpha : (\exists s \in \mathscr{S}) \, \alpha = \{\bar{p} \mid s, p \mid \bar{s}\}\},$$

$$\mathscr{D}' = \{d : \Omega \to \mathcal{O}\!\mathcal{U}'\}.$$

Every admissible rule is equivalent to

$$d'(\omega) = \{\bar{p} \mid s_\omega, p \mid \bar{s}_\omega\} \quad \text{where} \quad (\forall s \in \mathscr{S}) \, s_\omega \gtrsim_\omega s.$$

The rule d' is a CP version of the well-known maximum *a posteriori* decision rule of selecting the most probable state.

4. Extensions

Theorem 12, by enabling us to determine a partial ordering $\gtrsim_{\mathscr{D}}$ in terms of $\gtrsim_{\mathscr{P}}$, only leads to an optimal decision in special cases. Furthermore, except in special cases, we are unable to compare any two given rules. One remedy for this is to assume the existence of a CP relation $\gtrsim$ on all of $\Omega \times \mathscr{S}$. After straightforward modification of the axioms it can be shown that $(\mathscr{D}, \gtrsim_{\mathscr{D}})$ can now be embedded in $(\mathscr{P}, \gtrsim_{\mathscr{P}})$. Hence $\gtrsim_{\mathscr{P}}$ completely determines a complete order $\gtrsim_{\mathscr{D}}$. Another direction for the development of decision-making in a comparative probability context is to replace the structural axioms DCP5 and DCP6. For example, if we follow the line of utility theory (see Section VIIIB), then we might attempt to embed $(\mathscr{D}, \gtrsim_{\mathscr{D}})$ in $(\Omega \times \mathscr{S}, \gtrsim)$ or, by adjoining an auxiliary random experiment Λ, in $(\Omega \times \mathscr{S} \times \Lambda, \gtrsim)$ or even in $(\Lambda, \gtrsim_\Lambda)$, where

$$(\forall A, B \subseteq \Lambda)\, A \gtrsim_\Lambda B \Leftrightarrow (\Omega, \mathscr{S}, A) \gtrsim (\Omega, \mathscr{S}, B).$$

Some discussion of the second possibility can be found in Fine [18].

Finally, we may wish to augment the axioms of rationality. Two possible axioms, concerning continuity and behavior under randomization, are presented in the next section on a definition of expectation.

IIH. Expectation in Comparative Probability

The preceding analysis of decision-making in CP suggests a direction for a definition of a concept of expectation. In quantitative probability expectation is a continuous linear functional on the space of random variables. In utility-theory-based decision-making expectation enters in a representation of a preference pattern. If we adopt the latter viewpoint as basic, at least to the intuitive use of mathematical expectation as a measure of what it is reasonable to expect in an uncertain situation, then we can develop a definition of the expectation of acts as follows. Expectation $\mathscr{E}$ is to be an operator on acts yielding payoffs that are equivalent in value to the acts. With the notation introduced in the preceding section we can write:

E1. $\mathscr{E} : \mathcal{O} \to \mathscr{P}$.

E2. $(\forall p \in \mathscr{P})\, \mathscr{E}(\{p \mid \mathscr{S}\}) = p$.

E3. $\alpha_1 \gtrsim_\mathcal{O} \alpha_2 \Leftrightarrow \mathscr{E}(\alpha_1) \gtrsim_\mathscr{P} \mathscr{E}(\alpha_2)$.

If $(\mathcal{O}, \gtrsim_\mathcal{O})$ and $(\mathscr{P}, \gtrsim_\mathscr{P})$ satisfy DCP1–DCP6, then Theorem 11 assures us of the existence of a function $\mathscr{E}$ satisfying E1–E3.

To further structure the notion of expectation through a continuity property we require a

Definition. $\alpha_i \uparrow \alpha \Leftrightarrow (\forall i)\alpha \gtrsim_\mathcal{O} \alpha_{i+1} \gtrsim_\mathcal{O} \alpha_i,\, (\forall \alpha \succ_\mathcal{O} \beta)(\exists i)(\alpha \gtrsim_\mathcal{O} \alpha_i \succ_\mathcal{O} \beta)$.

E4. $\alpha_i \uparrow \alpha \Rightarrow \{\mathscr{E}(\alpha_i) \mid \mathscr{S}\} \uparrow \{\mathscr{E}(\alpha) \mid \mathscr{S}\}$.

The fifth and final axiom refers to the "linearity" of expectation as exhibited by its behavior with respect to randomization.

In keeping with the discussion of CP independence in Section IIF, we define

$$A \perp\!\!\!\perp \{B_1, ..., B_n\} \Leftrightarrow (\forall k)(\forall 1 \leqslant i_1 < \cdots < i_k \leqslant n)$$

$$C_j = \begin{cases} B_j & \text{if } j \in \{i_m\} \\ \bar{B}_j & \text{if } j \notin \{i_m\}, \end{cases}$$

$$A \perp\!\!\!\perp \bigcup_{j=1}^{n} C_j \,.$$

Equivalently we could require that the experiment having outcomes A or $\bar{A}$ be independent of the experiment with events in the smallest field of events generated from $\{B_1, ..., B_n\}$.

E5. $\quad \alpha = \{p_i \mid B_i\}, \quad \alpha' = \{p_i' \mid B_i\}, \quad \alpha'' = \{\mathscr{E}(\alpha) \mid A, \mathscr{E}(\alpha') \mid \bar{A}\},$

$\alpha''' = \{p_i \mid A \cap B_i, p_i' \mid \bar{A} \cap B_i\}, \quad A \perp\!\!\!\perp \{B_1, ..., B_n\} \Rightarrow \mathscr{E}(\alpha'') \approx_{\mathscr{P}} \mathscr{E}(\alpha''').$

The act α''' can be viewed as resulting from a randomization between acts α and α' based on the event A; if A occurs, then we perform α, whereas if $\bar{A}$ occurs, then we perform α'. Axiom E5 asserts that the expectation of a mixture of acts is equivalent to a mixture of their expectation.

The preceding definition of expectation $\mathscr{E}$ through E1–E5 contains the usual definition of (quantitative) mathematical expectation as a special case.

Theorem 13. If $\mathscr{P}$ is the set of real numbers, $\gtrsim_{\mathscr{P}}$ corresponds to numerical inequality, $(\mathscr{S}, \gtrsim)$ admits of an agreeing quantitative probability P,

$$\alpha_1 \gtrsim_\alpha \alpha_2 \Leftrightarrow E\alpha_1 = \int_{\mathscr{S}} \alpha_1(s) \, P(ds) \geqslant E\alpha_2 \,,$$

and $\mathscr{E}$ satisfies E1–E3, then $\mathscr{E}(\alpha) = E\alpha$, and $\mathscr{E}$ also satisfies E4 and E5.

II. Appendix: Proofs of Results

Theorem 1. If $\gtrsim$ satisfies C1–C4, then it admits of a quantitative probability P if and only if $\gtrsim$ also satisfies C5, where C5 can be taken to be either C5a or C5b.

Proof. A proof for C5a is available in Fine [20]. A proof for C5b is available in Fishburn [21]. ∎

Theorem 2.

(a)   C0–C5 $\not\Rightarrow$ C6,
(b)   C0–C4, C6 $\not\Rightarrow$ C5.

Proof.   (a)   A counterexample is provided in Subsection IIB3.
(b)   A counterexample is provided in Subsection IIB2.   ∎

Theorem 3.

(a)   C0–C6 $\not\Rightarrow$ C7,
(b)   C0–C4, C7 $\Rightarrow$ C5 and C6.

Proof.   (a)   A counterexample is provided by any $\gtrsim$ having a finitely but not countably additive agreeing P.
(b)   We first verify that C0–C4 , C7 $\Rightarrow$ C5.

(1)   Let $\Omega = \Omega_1 \cup \Omega_2$, where Ω_1 is atomless and Ω_2 contains only the atoms of Ω, $\Omega_2 = \{\omega_i\}$.

(2)   If $\mathscr{F}_i = \{F : F' \in \mathscr{F}, F = F' \cap \Omega_1\}$,   and $\Omega_1 > \varnothing$,   then by Villegas' theorem $\exists P_1$, $P_1 : \mathscr{F}_1 \to [0, 1]$ agrees with $\gtrsim$, and is countably additive. The case of $\Omega_1 \approx \varnothing$ can be treated using part of the remaining steps of this proof.

(3)   Define

$$\mathscr{C}_1 = \{H : H \in \mathscr{F}_1 , P_1(H) \text{ rational}\}$$

$$\mathscr{C}_2 = \{H : H \in \mathscr{F}_2 , \|H\| < \infty\} \cup \Omega_2$$

$$\mathscr{C} = \{F : F \cap \Omega_1 \in \mathscr{C}_1 \text{ and } F \cap \Omega_2 \in \mathscr{C}_2\}.$$

The quotient spaces $\mathscr{C}_1/\approx$ and $\mathscr{C}_2/\approx$ are countable. Hence $\mathscr{C}/\approx$ is countable.

(4)   From the countability of $\mathscr{C}/\approx$ and Theorem 1, $\exists P_2 : \mathscr{C} \to [0, 1]$, agreeing with $\gtrsim$ on $\mathscr{C}$.

(5)   Note that

$$H \in \mathscr{F}_2 \Rightarrow H = \bigcup_{j=1}^{\infty} w_{i_j} = \bigcup_{j=1}^{\infty} H_j \qquad \text{where} \quad H_j = \bigcup_{k=1}^{j} \omega_{i_k} \in \mathscr{C}_2 .$$

Hence

$$(\forall H \in \mathscr{F}_2)(\exists\{H_i\})\,(H_i \in \mathscr{C}_2 , H_i \uparrow H).$$

(6)   Assert that

$$(\forall H \in \mathscr{F}_1)(\exists\{H_i\})\left(H_i \in \mathscr{C}_1 , \bigcup_{i=1}^{\infty} H_i \approx H\right).$$

To verify, note that

$$H \in \mathscr{F}_1 \Rightarrow P_1(H) = y \in [0, 1] \quad \text{and} \quad (\forall y)(\exists\{x_i\}) \left(\sum_{i=1}^{n} x_i \uparrow y, \, x_i \text{ rational} \right).$$

Since the range of countably additive P_1 for $\mathscr{F}_1$ atomless is an interval,

$$(\forall x_i)(\exists G_i) \left(P_1(G_i) = x_i \right).$$

Choose $\{G_i\}$ pairwise disjoint, as is possible. Define

$$H_i = \bigcup_{j=1}^{i} G_j \, .$$

Note,

$$\lim_{i \to \infty} P_1(H_i) = \sum_{}^{\infty} x_i = y = P_1 \left(\bigcup_{}^{\infty} H_i \right),$$

where the last equality follows from P_1 countably additive. Hence

$$H_i \uparrow \bigcup_{}^{\infty} H_i \approx H,$$

as claimed.

(7)   Claim that

$$(\forall F \in \mathscr{F})(\exists\{F_i\}) \left(F_i \in \mathscr{C}, F_i \uparrow \bigcup_{j=1}^{\infty} F_j \approx F \right).$$

To verify, note that

$$F = (F \cap \Omega_1) \cup (F \cap \Omega_2)$$

and apply (5) and (6) in each of the disjoint parts of F. Hence,

$$(\exists H_i\uparrow, K_i\uparrow) \left(\bigcup_{}^{\infty} H_i \approx F \cap \Omega_1 , \bigcup_{}^{\infty} K_i = F \cap \Omega_2 \right).$$

If we define

$$F_i = H_i \cup K_i \, ,$$

then $\{F_i\}$ verifies our claim.

(8)   We define

$$P : \mathscr{F} \to R^1$$

through

$$P(F) = \sup\{x : H \in \mathscr{C}, H \precsim F, x = P_2(H)\}.$$

(9)   To verify that P agrees with $\gtrsim$ we first show that $P(F) > P(G) \Rightarrow F \succ G$. It follows from (8) that

$$P(F) > P(G) \Rightarrow (\exists H \in \mathscr{C}) F \gtrsim H \succ G$$

Hence, $F \succ G$.

(10)   To complete the proof that P agrees with $\gtrsim$ we need to verify that

$$F \succ G \Rightarrow P(F) > P(G).$$

This is trivial if both $F, G \in \mathscr{C}$. Hence assume, say, $F \notin \mathscr{C}$. By (7)

$$(\exists F_i \uparrow) \left(\bigcup_{i}^{\infty} F_i \approx F \right).$$

If $(\forall n)(F_n \lesssim G)$, then by monotone continuity

$$\bigcup^{\infty} F_i \lesssim G \prec F.$$

However, this contradicts $\bigcup^{\infty} F_i \approx F$. Hence

$$(\exists n) F \gtrsim F_n \succ G.$$

Furthermore, by $F \notin \mathscr{C}(\exists m > n)$ $F \gtrsim F_m \succ F_n \succ G$. It follows that

$$P(F) \geqslant P(F_m) > P(F_n) \geqslant P(G).$$

Hence $P(F) > P(G)$ as required for the first part of this proof. The case of $G \notin \mathscr{C}$ can be treated similarly.

To verify that C0–C4, C7 $\Rightarrow$ C6 we require Villegas' theorem as given in Section IID. The immediate consequences of this theorem are, as pointed out by Villegas, as follows. If $(\Omega, \mathscr{F}, \gtrsim)$ satisfies C0–C4 and C7, then

$$\Omega = \Omega_a \cup \Omega_c, \qquad \Omega_a \cap \Omega_c = \varnothing,$$

where Ω_a is a countable collection of atoms, which, if nonempty, can be enumerated in nonincreasing order, $\Omega_a = \{\omega_i\}$, $\omega_i \gtrsim \omega_{i+1}$, and Ω_c, if nonempty, is atomless and supports a unique countably additive P agreeing with $\gtrsim$ restricted to Ω_c. In the remainder of the proof we assume $\Omega_a \succ \varnothing$, $\Omega_c \succ \varnothing$; if either are null-equivalent, then the proof can be simplified in obvious ways.

Assume, to the contrary of our expectations, that there is some $A \succ \varnothing$ for which there is an infinite-length upper scale $\{A_i\}$. We assert that for any $B \prec A$, $\{A_{2i} - B\}$ is also an infinite-length upper scale for A. To verify this note

$$A_{2i+2} - B > A_{2i+1} \gtrsim A_{2i+1} - B > A_{2i} \gtrsim A_{2i} - B,$$

and that the gap from $A_{2i} - B$ to $A_{2i+2} - B$ exceeds that from A_{2i} to A_{2i+1}. To choose B appropriately we note that by C7,

$$(\exists N) \bigcup_{i>N} \omega_i \prec A.$$

Take $B = \bigcup_{i>N} \omega_i$. Hence, if there is any infinite-length upper scale for some nonnull A, then there is also an infinite-length upper scale for A no term of which contains an atom in B. With this in mind, we now assume that our postulated infinite upper scale for A satisfies

$$(\forall n) A_n \cap \left(\bigcup_{i>N} \omega_i \right) = \varnothing.$$

For convenience, we denote

$$B_n = A_n \cap \Omega_a, \qquad C_n = A_n \cap \Omega_c.$$

Remembering that B_n is a subset of the finite set $\{\omega_1, ..., \omega_N\}$ we see that there are only finitely many such subsets. At least one subset must therefore occur infinitely often. Let B' be a maximal (in the sense of $\gtrsim$) subset of $\{\omega_1, ..., \omega_N\}$ that occurs infinitely often in $\{A_n\}$ and denote the indices of occurrence by $\{i_j\}$,

$$A_{i_j} \cap \Omega_a = B'.$$

From the definition of an upper scale

$$A_{i_{j+1}} = B' \cup C_{i_{j+1}} \gtrsim A_{i_{j+1}} > A_{i_j} = B' \cup C_{i_j},$$

and

$$(\forall i_j)(\exists D, E) D \gtrsim A, \qquad E \gtrsim A_{i_j}, \qquad D \cap E = \varnothing, \qquad A_{i_{j+1}} \gtrsim D \cup E.$$

Noting that $C_{i_j} \subset \Omega_c$, atomless, we may without loss of generality take $C_{i_j} \subset C_{i_{j+1}}$. If $C_{i_{j+1}} - C_{i_j} \prec A$, then take $E = A_{i_j}$ and $A > D = C_{i_{j+1}} - C_{i_j}$, to obtain $A_{i_{j+1}} = D \cup E$, in contradiction to the definition of an upper scale. Thus $C_{i_{j+1}} - C_{i_j} \gtrsim A$.

Now choose $A' \subset \Omega_c$, such that

$$A \gtrsim A' > \varnothing$$

to obtain

$$P(C_{i_{j+1}}) \geqslant P(C_{i_j}) + P(A').$$

As $\{i_j\}$ is assumed infinite, we reach the contradiction that $P(C_{i_j})$ diverges to infinity. Hence there are no infinite-length upper scales for nonnull events A. This verifies C6. ∎

Theorem 4: If $\gtrsim$ satisfies C0–C5, then there exists P agreeing with $\gtrsim$, and there exists a function G of two variables such that

$$A \cap B = \varnothing \Rightarrow P(A \cup B) = G(P(A), P(B)).$$

Furthermore, $G(x, y)$ is symmetric $[G(x, y) = G(y, x)]$, strictly increasing in x, $G(x, P(\phi)) = x$, and associative $[G(G(x, y), z) = G(x, G(y, z))]$.

Proof. The existence of P is immediate from Theorem 1. The existence of G follows if we can demonstrate that $A \cap B = C \cap D = \varnothing$, $P(A) = P(C)$, and $P(B) = P(D) \Rightarrow P(A \cup B) = P(C \cup D)$. To see this, first note that $P(E) = P(F) \Rightarrow E \approx F$. It is an easy consequence of (d) in Section IIB that $A \approx B$, $C \approx D$, $A \cap C = \varnothing$, and $B \cap D = \varnothing \Rightarrow A \cup C \approx B \cup D$. Hence under our hypothesis $A \cup B \approx C \cup D$, and it follow from the definition of agreeing P that $P(A \cup B) = P(C \cup D)$, as desired.

That G is symmetric is immediate from $A \cup B = B \cup A$ when we let $x = P(A)$ and $y = P(B)$.

To verify that G is increasing let $A \cap C = \varnothing$, $B \cap D = \varnothing$, $P(A) = x$, $P(B) = x'$, $P(C) = y$, and $P(D) = y'$ with $x' > x$ and $y' \geqslant y$. Invoking properties (d) and (e) of Section IIB and the fact that P agrees with $\gtrsim$ leads to $B \cup D > A \cup C$. Hence $P(B \cup D) = G(x', y') > P(A \cup C) = G(x, y)$.

If $A \cap B = \varnothing$, $P(A) = x$, and $P(B) = 0$, then $B \lesssim \varnothing$ and by C4, $A \cup B \lesssim A \cup \varnothing = A$. However, $B \gtrsim \varnothing \Rightarrow A \cup B \gtrsim A$. Thus $A \cup B \approx A$ and $P(A \cup B) = G(x, P(\phi)) = P(A) = x$.

Finally, to verify associativity assume $A \cap (B \cup C) = \varnothing$, $B \cap C = \varnothing$, $P(A) = x$, $P(B) = y$, and $P(C) = z$. Then $P(A \cup (B \cup C)) = G(x, G(y, z))$ Note that $A \cap (B \cup C) = \varnothing$, $B \cap C = \varnothing \Rightarrow C \cap (A \cup B) = \varnothing$, and $A \cap B = \varnothing$. Hence,

$$P((A \cup B) \cup C) = G(G(x, y), z).$$

However, by the associativity of union $A \cup (B \cup C) = (A \cup B) \cup C$, and therefore $G(x, G(y, z)) = G(G(x, y), z)$. ∎

Theorem 5. If a finitely additive P satisfies $(***)$ and almost agrees with $\succsim$ satisfying C1–C4, then P agrees with $\succsim$ if and only if $\succsim$ satisfies C5.

Proof. The "only if" part is trivial. To verify the "if" part assume P' agrees with $\succsim$. From Savage's theorem we know that there exists P additive and almost agreeing with $\succsim$. Either P agrees with $\succsim$, in which case there is nothing to be proved, or it does not. If P does not agree with $\succsim$, then there exist F and G such that $F \succ G$ but $P(F) = P(G)$. Either $P(F \cup G) < P(\Omega)$ or $P(F \cup G) = P(\Omega)$.

If $P(F \cup G) < P(\Omega)$, then $(\forall 0 < \alpha \leqslant 1)$ $(\exists H_\alpha \subseteq \overline{F \cup G})$ $(P(H_\alpha) = \alpha P(\overline{F \cup G}) > 0)$.

Consider the real intervals

$$(0 < \alpha \lesssim 1) \qquad I_\alpha = (P'(G \cup H_\alpha), P'(F \cup H_\alpha)).$$

We claim that for $\alpha \neq \beta$,

$$I_\alpha \cap I_\beta = \varnothing, \qquad I_\alpha \neq \varnothing.$$

To verify this claim say $\alpha > \beta$, and hence,

$$P(G \cup H_\alpha) = P(G) + \alpha P(\overline{F \cup G}) > P(G) + \beta P(\overline{F \cup G})$$
$$= P(F) + \beta P(\overline{F \cup G})$$
$$= P(F \cup H_\beta).$$

Since P almost agrees with $\succsim$, $G \cup H_\alpha \succ F \cup H_\beta$. Hence, in terms of the agreeing P' we find that

$$\alpha > \beta \Rightarrow P'(G \cup H_\beta) < P'(F \cup H_\beta) < P'(G \cup H_\alpha) < P'(F \cup H_\alpha).$$

Therefore, corresponding to each α there is a nonvoid I_α, with disjoint I_α and I_β for $\alpha \neq \beta$. However, there are uncountably many values of α but only countably many disjoint, nonvoid, open intervals, and we have reached a contradiction. Hence, if $P(F \cup G) < P(\Omega)$, then $P(F) = P(G) \Rightarrow F \approx G$.

To complete the proof we need to treat the case of $P(F \cup G) = P(\Omega)$, where $F \succ G$, and $P(F) = P(G)$. This case can be dealt with by reducing it to the previous case as follows. Replace F and G by $A = F - (F \cap G)$ and $B = G - (F \cap G)$, respectively, where now $P(A) = P(B)$ and $A \succ B$. If $P(A \cup B) < P(\Omega)$, then we have just proved that $A \approx B$ and this is a desired contradiction. If $P(A \cup B) = P(\Omega)$, then by Savage's theorem there exist $A' \subseteq A$ and $B' \subseteq B$ such that $P(A') =$

$P(B') = \frac{1}{2}P(A)$. Since $(A - A') \cup A' > (B - B') \cup B'$, at least one of $(A - A')$ or A' is more probable than one of $(B - B')$ or B', say $A' > B'$. However, $P(A' \cup B') = \frac{1}{2}P(A \cup B) < P(\Omega)$. Hence, by our preceding discussion $A' \approx B'$, and we again reach the desired contradiction.

Thus, we have shown that if $P(F) = P(G)$, then $F \approx G$; the almost agreeing, additive P in fact agrees with $\gtrsim$. ∎

Theorem 6. (C1–C5, S) $\Rightarrow$ (C1–C4, C6, L), but the converse implication is not valid.

Proof. From the corollary to Theorem 5 we see that C1–C5 and S imply the existence of an additive P satisfying ($***$) and agreeing with $\gtrsim$. Hence if $\Omega > A > \varnothing$, then $\infty > P(A) > 0$. Furthermore $A > \varnothing \Rightarrow \bar{A} < \Omega$ and hence that $P(\bar{A}) < \infty$. Thus $P(\Omega) = P(A) + P(\bar{A}) < \infty$; the Archimedean property C6 is now immediate. To verify L note that if $\rho = P(C)/P(A \cup B)$ and $\sigma = P(D)/P(A \cup B)$, where $A \cap B = \varnothing$, $A \gtrsim C$, and $C \gtrsim D$, then by ($***$) there exists $C' \subseteq A \cup B$ such that $P(C') = \rho P(A \cup B) = P(C)$. Hence $C' \approx C$ and similarly we can find $D' \approx D, D' \cap C' = \varnothing$. This verifies that (C1–C5, S) $\Rightarrow$ (C1–C4, C6, L) To confirm the falsity of the reverse implication it suffices to take as a counterexample any two point space Ω. ∎

Theorem 7. If $\gtrsim$ admits of an agreeing finitely additive P, then P is countably additive if and only if $\gtrsim$ satisfies C8.

Proof. To verify "only if" assume that P is countably additive and that, contrary to our expectation,

$$\exists \{B_i\} \, (B_i \downarrow \varnothing, \exists A_0 \in \bigcap_{i=1}^{\infty} \{A : \varnothing < A \lesssim B_i\}).$$

Hence $(\forall i) \, P(B_i) \geqslant P(A_0) > P(\varnothing) = 0$. However, by the continuity property of countably additive P, $\lim_{i \to \infty} P(B_i) = 0$, and this contradicts $P(A_0) > 0$. Thus $\{A : \varnothing < A \lesssim B_i\} = \varnothing$ as claimed.

To verify "if" assume that, contrary to our expectations, $\exists \{B_i\} \, B_i \downarrow \varnothing$, $\bigcap_i \{A : \varnothing < A \lesssim B_i\} = \varnothing$ yet $\lim_{i \to \infty} P(B_i) = \epsilon > 0$. Hence there is no A for which $\epsilon > P(A) > 0$; if there were such an A, then $(\forall i)$ $(\varnothing < A \lesssim B_i)$ in contradiction to $\bigcap_i \{A : \varnothing < A \lesssim B_i\} = \varnothing$. It follows from $B \downarrow \varnothing$, and $P(B_i) \downarrow \epsilon$ that if for some n, $\delta \leqslant \epsilon$, $\epsilon + \delta > P(B_n) > \epsilon$, then $(\exists k > n)$ $(\delta > P(B_n - B_k) > 0)$. This, however, contradicts $(\nexists A)$ $(\epsilon > P(A) > 0)$. Hence, either $P(B_i) \geqslant 2\epsilon$ or $P(B_i) = \epsilon$, and it follows from $P(B_i) \downarrow \epsilon$ that $(\exists j)$ $(P(B_j) = \epsilon)$.

However, this shows that $(\forall i)\,(B_j \in \{A : \varnothing \prec A \precsim B_i\})$, in contradiction to $\bigcap \{A : \varnothing \prec A \precsim B_i\} = \varnothing$. The contradiction can only be avoided by rejecting the tentative hypothesis that $\epsilon > 0$. Hence, P is continuous, and therefore countably additive. ∎

Theorem 8. If $\succsim^*$ satisfies QCC1, QCC2, and QCC5, then there exists some agreeing function P. There exists a function G of two variables such that

(1) $P((A \cap B)/C) = G(P(B/C), P(A/B \cap C))$,

(2) $G(x, y) = G(y, x)$ (symmetry),

(3) $G(x, y)$ is increasing in x for $y > P(\varnothing/B)$,

(4) $G(x, G(y, z)) = G(G(x, y), z)$ (associativity),

(5) $G(P(\Omega/B), y) = y$,

(6) $G(P(\varnothing/B), y) = P(\varnothing/B)$,

if and only if $\succsim^*$ also satisfies QCC7.

Proof. Theorem 1 establishes the existence of some agreeing P. To verify "if" assume that $\succsim^*$ satisfies QCC7.

(1) Assume A, B, C, A', B', and C' with

$$P(B/C) = P(B'/C') = x, \qquad P(A/(B \cap C)) = P(A'/(B' \cap C')) = y,$$

$$P((A \cap B)/C) = z.$$

G fails to exist if and only if there exist A', B', and C' such that the above holds, yet $P(A' \cap B'/C') = z' \neq z$. To see that this is impossible note that since P agrees with $\succsim^* : B/C \approx^* B'/C'$, $A/(B \cap C) \approx^* A'/(B' \cap C')$. Recalling that $B/C \approx^* B'/C' \Rightarrow B/C \succsim^* B'/C'$, $B/C \succsim^* B'/C'$, and invoking QCC7(a) we see that $(A' \cap B')/C' \approx (A \cap B)/C$. Hence, since P agrees with $\succsim^*$, $z' = z$ and G is well defined.

(2) Assume that $P(A/B \cap C) = P(B'/C') = x$ and $P(B/C) = P(A'/B' \cap C') = y$. Then $P(A \cap B/C) = G(x, y)$ and $P(A' \cap B'/C') = G(y, x)$. P agrees with $\succsim^*$ implies that $A/B \cap C \approx^* B'/C'$ and $B/C \approx^* A'/B' \cap C'$. Invoking QCC7(b) we see that $A \cap B/C \approx^* A' \cap B'/C'$. Hence $P(A \cap B/C) = P(A' \cap B'/C')$, and G is symmetric.

(3) $P(A/B \cap C) = x$, $P(B/C) = y$, $P(A'/B' \cap C') = x'$, $P(B'/C') = y'$, $P(A \cap B/C) = z$, and $P(A' \cap B'/C') = z'$ with $x' > x$ and $y' \geqslant y > P(\varnothing/B)$. We claim that $z' > z$. Note that $A'/B' \cap C' \succ^* A/B \cap C$ and $B'/C' \succsim^* B/C \succ^* \varnothing/B$. Invoking QCC7(c), we see that $A' \cap B'/C' \succ^* A \cap B/C$. Since P agrees with $\succsim^*$ it follows that $z' = P(A' \cap B'/C') > P(A \cap B/C) = z$, and G is increasing.

(4) Assume that $x = P(A/B \cap C \cap D)$, $y = P(B/C \cap D)$, and $z = P(C/D)$. By (1)

$$
\begin{aligned}
P(A \cap B \cap C/D) &= G(P(A \cap B/C \cap D), P(C/D)) \\
&= G(G(P(A/B \cap C \cap D), P(B/C \cap D)), P(C/D)) \\
&= G(G(x, y), z),
\end{aligned}
$$

and

$$
\begin{aligned}
P(A \cap B \cap C/D) &= G(P(A/B \cap C \cap D), P(B \cap C/D)) \\
&= G(P(A/B \cap C \cap D), G(P(B/C \cap D), P(C/D))) \\
&= G(x, G(y, z)).
\end{aligned}
$$

Hence G is associative.

(5)   Assume $P(A/B) = y$. Then

$$
y = P(A \cap \Omega/B) = G(P(\Omega/B), P(A/B \cap \Omega)) = G(P(\Omega/B), y).
$$

(6)   If $y = P(A/B)$, then

$$
P(\varnothing/B) = P(\varnothing \cap A/B) = G(P(\varnothing/B \cap A), P(A/B)).
$$

Hence

$$
P(\varnothing/B) = G(P(\varnothing/B), y).
$$

To verify "only if" we assume the existence of G having properties (1), (2), and (3) and show that $\gtrsim^*$ satisfies QCC7. The details are straightforward and are omitted. ∎

Theorem 9.   If $\gtrsim^*$ satisfies QCC1–QCC7, $\Omega \in \mathscr{G}$, then there exist P agreeing with $\gtrsim^*$ and real-valued functions F and G of two variables such that if $A/\Omega >^* \varnothing/\Omega$, then

$$
P(B/A) = F(G(P(A/B), P(B/\Omega)), P(A/\Omega)).
$$

Furthermore, G has properties (1)–(6) listed in Theorem 8; $F(x, y)$ is increasing in x and decreasing in y.

Proof.   From the corollary in Subsection IIE2 we have the existence of agreeing P and of G such that

$$
P(A \cap B/C) = G(P(A/B \cap C), P(B/C)) = G(P(B/A \cap C), P(A/C)),
$$

with G having properties (1)–(6) of Theorem 8. Hence, letting $C = \Omega$, we have the identity

$$
G(P(A/B), P(B/\Omega)) = G(P(B/A), P(A/\Omega)).
$$

Since $P(A/\Omega) > P(\varnothing/\Omega)$ we have from the monotonicity of G that it can be inverted: There exists F such that

$$G(x, y) = z, \qquad y > P(\varnothing/\Omega) \Rightarrow x = F(z, y).$$

Therefore,

$$P(B/A) = F\big(G(P(A/B), P(B/\Omega)), P(A/\Omega)\big).$$

This verifies the existence of the asserted representation and the properties of G. From the monotonicity of G it follows that $F(z, y)$ is increasing in z and decreasing in y. ∎

Theorem 10. If $\succsim$ satisfies C1–C5, and $\perp\!\!\!\perp$ satisfies I1 and I5, then there exist P agreeing with $\succsim$ and G such that

$$A \perp\!\!\!\perp B \Rightarrow P(A \cap B) = G(P(A), P(B)).$$

Furthermore $G(x, y)$ is symmetric and increasing in x for $y > P(\phi)$.

Proof. The existence of P was established by Theorem 1. By the existence of G we mean that whenever $P(A) = x$, $P(B) = y$, $A \perp\!\!\!\perp B$, then $P(A \cap B)$ has a fixed value $G(x, y)$; that is, G is single-valued. Assume to the contrary that there exist C and D such that $A \approx C$, $B \approx D$, and $C \perp\!\!\!\perp D$, yet $P(C \cap D) \neq P(A \cap B)$. Hence $C \cap D$ is not equivalent to $A \cap B$. However, this contradicts the immediate consequence of I5 that

$$A \approx C, \qquad B \approx D, \qquad A \perp\!\!\!\perp B, \qquad \text{and} \qquad C \perp\!\!\!\perp D \Rightarrow A \cap B \approx C \cap D.$$

The symmetry of G is immediate from I1. To verify that G is increasing in x for $y > P(\varnothing)$, consider $A \perp\!\!\!\perp B$, $C \perp\!\!\!\perp D$, $P(A) = x' > P(C) = x$, and $P(B) = y' \geqslant P(D) = y > P(\varnothing)$. From I5, $A \cap B \succ C \cap D$ and $P(A \cap B) > P(C \cap D)$. Hence $G(x', y') > G(x, y)$. ∎

Lemma 1. G as defined in Theorem 10 is associative, when all terms are defined, if and only if $\perp\!\!\!\perp$ satisfies I6.

Proof. For convenience, let

$$P(A) = P(A') = x, \qquad P(B) = P(B') = y, \qquad P(C) = P(C') = z.$$

If G is associative, then it is immediate that the implication asserted by I6, that $A \cap B \cap C \approx A' \cap B' \cap C'$, holds.

If there are events $A \approx A'$, $B \approx B'$, and $C \approx C'$ such that $B \perp\!\!\!\perp C$, $A \perp\!\!\!\perp B \cap C$, $A' \perp\!\!\!\perp B'$, and $C' \perp\!\!\!\perp A' \cap B'$, then $G(x, y)$, $G(y, z)$, $G(x, G(y, z))$ and $G(G(x, y), z)$ are all defined. If I6 holds, then

$P(A \cap B \cap C) = P(A' \cap B' \cap C')$, and hence, G is associative. Note though that we cannot assert that $G(x, y)$ and $G(G(x, y), z)$ defined implies that $G(G(x, y), z) = G(x, G(y, z))$. The terms $G(y, z)$ or $G(x, G(y, z))$ need not be defined; for example, there may not exist

$$B', \quad C', \quad P(B') = y, \quad P(C') = z, \quad \text{and} \quad B' \perp\!\!\!\perp C'. \quad \blacksquare$$

Lemma 2. If DCP1, DCP3, DCP4, $\{E_i\}$ is any finite partition of $\mathscr{S}$, and

$$\alpha = \{p_i \mid E_i\}, \qquad \alpha' = \{p_i' \mid E_i\},$$

then

(a) $(\forall i)\, p_i \gtrsim_{\mathscr{P}} p_i' \Rightarrow \alpha \gtrsim_\alpha \alpha'$,

(b) $(\forall i)\, p_i \gtrsim_{\mathscr{P}} p_i', (\exists j)\, (p_j >_{\mathscr{P}} p_j', E_j > \varnothing) \Rightarrow \alpha >_\alpha \alpha'$.

Proof. Recursively define $\{\alpha_k''\}$ by

$$\alpha_0'' = \alpha',$$

$$\alpha_{k+1}''(s) = \begin{cases} p_{k+1} & \text{if } s \in E_{k+1} \\ \alpha_k''(s) & \text{if } s \notin E_{k+1}. \end{cases}$$

Since α_{k+1}'' and α_k'' agree on $\bar{E}_{k+1}$ it follows from DCP4 that

$$\alpha_{k+1}'' \gtrsim_\alpha \alpha_k'' \Leftrightarrow \{p_{k+1} \mid \mathscr{S}\} \gtrsim_\alpha \{p_{k+1}' \mid E_{k+1}, p_{k+1} \mid \bar{E}_{k+1}\}.$$

Invoking the hypothesis that $p_{k+1} \gtrsim_{\mathscr{P}} p_{k+1}'$ and DCP3 yields

$$p_{k+1} \gtrsim_{\mathscr{P}} p_{k+1}' \Rightarrow \alpha_{k+1}'' \gtrsim_\alpha \alpha_k'',$$

$$p_{k+1} >_{\mathscr{P}} p_{k+1}', E_{k+1} > \varnothing \Rightarrow \alpha_{k+1}'' >_\alpha \alpha_k''.$$

Since $\{E_i\}$ is a finite partition of $\mathscr{S}$ ($\mathscr{S}$ is a finite set), there is an n for which $\alpha_n'' = \alpha$. From the transitivity of $\gtrsim_\alpha$ given by DCP1, we conclude with $\alpha \gtrsim_\alpha \alpha'$, and if

$$(\exists j)\, p_j >_{\mathscr{P}} p_j', \qquad E_j > \varnothing,$$

then $\alpha >_\alpha \alpha'$. $\quad \blacksquare$

Theorem 11. If DCP1, DCP3, DCP4, DCP6, then

$$(\exists \pi : \mathscr{A} \to \mathscr{P})\, \alpha \gtrsim_\alpha \alpha' \Leftrightarrow \pi(\alpha) \gtrsim_{\mathscr{P}} \pi(\alpha').$$

Proof. The proof proceeds by a backwards induction on the number n of sets in the partition $\{E_i\}$ of $\mathscr{S}$ over which

$$\alpha = \{p_i \mid E_i\}.$$

By DCP6,

$$(\forall p_{n-1}, p_n \in \mathscr{P})(\forall E_{n-1}, E_n)(\exists p'_{n-1} \in \mathscr{P})$$

$$\{p'_{n-1} \mid E_{n-1} \cup E_n, p_{n-1} \mid \overline{E_{n-1} \cup E_n}\} \sim_\alpha \{p_n \mid E_n, p_{n-1} \mid \bar{E}_n\}.$$

Define

$$\alpha_{n-1}(s) = \begin{cases} p'_{n-1} & \text{if } s \in E_{n-1} \cup E_n \\ p_j & \text{if } s \in E_j \end{cases} \quad \text{for } j < n-1.$$

Note that α_{n-1} and α agree on $E_{n-1} \cup E_n$. Hence by DCP4,

$$\alpha_{n-1} \sim_\alpha \alpha_n \Leftrightarrow \{p'_{n-1} \mid E_n \cup E_{n-1}, p_{n-1} \mid \overline{E_n \cup E_{n-1}}\} \sim_\alpha \{p_n \mid E_n, p_{n-1} \mid \bar{E}_n\},$$

which was the defining property of p_{n-1}. Thus for every act defined on an n-fold partition of $\mathscr{S}$ there is an equivalent act defined on an $(n-1)$-fold partition of $\mathscr{S}$. Repeating this process and invoking the transitivity of $\gtrsim_\alpha$ assured by DCP1, we conclude that

$$(\forall \alpha)(\exists p_\alpha \in \mathscr{P}) \; \alpha \sim_\alpha \{p_\alpha \mid \mathscr{S}\}.$$

Let $\pi(\alpha) = p_\alpha$. Noting that $\mathscr{S} > \varnothing$ and invoking DCP3 yields

$$p \gtrsim_\mathscr{P} q \Leftrightarrow \{p \mid \mathscr{S}\} \gtrsim_\alpha \{q \mid \mathscr{S}\}.$$

Hence,

$$\alpha \gtrsim_\alpha \alpha' \Leftrightarrow \pi(\alpha) \gtrsim_\mathscr{P} \pi(\alpha'). \quad \blacksquare$$

Theorem 12. If each $\gtrsim_\alpha^\omega \in \{\gtrsim_\alpha^\omega, \omega \in \Omega\}$ satisfies DCP1, DCP3, DCP4, DCP6, and if DCP7 holds, then

(a) $(\forall \omega \in \Omega)(\exists \pi_\omega : \mathcal{O} \to \mathscr{P}) \; \alpha \gtrsim_\alpha^\omega \alpha' \Leftrightarrow \pi_\omega(\alpha) \gtrsim_\mathscr{P} \pi_\omega(\alpha')$

(b) $(\forall \omega \in \Omega) \; \pi_\omega(d_1(\omega)) \gtrsim_\mathscr{P} \pi_\omega(d_2(\omega)) \Rightarrow d_1 \gtrsim_\mathscr{D} d_2$.

(c) $(\forall \omega \in \Omega) \left(\pi_\omega(d_1(\omega)) \gtrsim_\mathscr{P} \pi_\omega(d_2(\omega))\right), (\exists \omega) \left(\pi_\omega(d_1(\omega)) >_\mathscr{P} \pi_\omega(d_2(\omega))\right)$
$\Rightarrow d_1 >_\mathscr{D} d_2$.

Proof. Part (a) merely restates Theorem 11. Parts (b) and (c), are immediate consequences of (a) and DCP7. $\quad \blacksquare$

Theorem 13. If $\mathscr{P}$ is the set of real numbers, $\gtrsim_\mathscr{P}$ corresponds to numerical inequality, $(\mathscr{S}, \gtrsim)$ admits of an agreeing quantitative probability P,

$$\alpha_1 \gtrsim_\alpha \alpha_2 \Leftrightarrow E\alpha_1 = \int_\mathscr{S} \alpha_1(s) \, P(ds) \gtrsim E\alpha_2,$$

and $\mathscr{E}$ satisfies E1–E3, then $\mathscr{E}(\alpha) = E\alpha$, and $\mathscr{E}$ also satisfies E4 and E5.

Proof. Since $\mathscr{E}(\alpha)$, by E3, and $E\alpha$, by hypothesis, both agree with the complete order $\succsim_\alpha$, there exists a strictly increasing function ϕ such that $\mathscr{E}(\alpha) = \phi(E\alpha)$. By $E\alpha \in \mathscr{P}$ and E2 we have that

$$\mathscr{E}(E\alpha \mid \mathscr{S}) = E\alpha.$$

Hence, taking $\alpha = \{x \mid \mathscr{S}\}$ we have that

$$E\alpha = x, \qquad \mathscr{E}(\alpha) = x.$$

Thus $\phi(x) = x$ and $\mathscr{E}(\alpha) = E\alpha$, as claimed. The verification of E4 and E5 for $E(\alpha)$ is immediate as they are restatements of well-known properties of $E\alpha$. ∎

References

1. M. Kaplan, Private Communication, Doctoral research, Cornell Univ., Ithaca, New York, 1971.
2. G. Debreu, Representation of a Preference Order by a Numerical Function, in *Decision Processes* (R. Thrall, C. Coombs, and R. Davis, eds.), p. 164. New York: Wiley, 1954.
3. J. Kelley, *General Topology*, pp. 46–48, 57, 58. Princeton, New Jersey: Van Nostrand-Reinhold, 1955.
4. R. D. Luce, Sufficient Conditions for the Existence of a Finitely Additive Probability Measure, *Ann. Math. Statist.* **38**, 780–786, 1967.
5. C. Villegas, On Qualitative Probability σ-Algebras, *Ann. Math. Statist.* **35**, 1787–1796, 1964.
6. J. Aczel, *Lectures on Functional Equations*, pp. 253–273. New York: Academic Press, 1966.
7. C. Kraft, J. Pratt, and A. Seidenberg, Intuitive Probability on Finite Sets, *Ann. Math. Statist.* **30**, 408–419, 1959.
8. D. Scott, Measurement Structures and Linear Inequalities, *J. Mathematical Psychology* **1**, 233–247, 1964.
9. L. J Savage, *Foundations of Statistics*, pp. 34–36. New York: Wiley, 1954.
10. P. C. Fishburn, Weak Qualitative Probability on Finite Sets, *Ann. Math. Statist.* **40**, 2118–2126, 1969.
11. J. M. Keynes, *A Treatise on Probability*. New York: Harper, 1962.
12. B. Koopman, The Bases of Probability, *Bull. Amer. Math. Soc.* **46**, 763–774, 1940.
13. Z. Domotor, *Probabilistic Relational Structures and Their Applications*. Tech. Rep. No 144, pp. 73–90. Inst. for Math. Studies in the Social Sci., Stanford Univ., Stanford, California, May 1969.
14. R. D. Luce, On the Numerical Representation of Qualitative Conditional Probability, *Ann. Math. Statist.* **39**, 481–491, 1968.

15. A. Renyi, *Foundations of Probability*, p. 38. San Francisco, California: Holden-Day, 1970.
16. A. Kolmogorov, *Foundations of the Theory of Probability*, p. 9. Bronx, New York: Chelsea, 1956.
17. M. Kaplan, Independence in Comparative Probability. M.S. Thesis, Cornell Univ., Ithaca, New York, 1970.
18. T. Fine, Rational Decision Making with Comparative Probability, *Proc. IEEE Conf. on Decision and Control, Miami Beach, 1971*, pp. 355–356. Gainesville: Dept. of Elec. Eng., Univ. of Florida, December 1971.
19. T. Ferguson, *Mathematical Statistics*, p. 54. New York: Academic Press, 1967.
20. T. Fine, A Note on the Existence of Quantitative Probability, *Ann. Math. Statist.* **42**, 1182–1186, 1971.
21. P. C. Fishburn, *Utility Theory for Decision Making*, pp. 27–29. New York: Wiley, 1970.

III

Axiomatic Quantitative Probability

IIIA. Introduction

Our concern is a critical discussion of axiomatic quantitative probability, primarily as developed by Kolmogorov [1] in 1929 and since widely accepted. We discuss events, probability, independence, and conditional probability, although not other important but, in our view, less basic concepts such as expectation and random variables. The Kolmogorov setup for probability consists of a probability space $(\Omega, \mathscr{F}, P)$ having as components a sample space, Ω; a σ-field (also called a σ-algebra) $\mathscr{F}$ of selected subsets of Ω; and a probability measure or assignment, P. The sample space Ω has elements ω called the elementary events. The σ-field of subsets of Ω, $\mathscr{F}$, has the following three properties.

(1) $\Omega \in \mathscr{F}$.
(2) If $F \in \mathscr{F}$, then $\bar{F} \in \mathscr{F}$ (closure under complementation).
(3) If for countably many i, $F_i \in \mathscr{F}$, then $\bigcup_i F_i \in \mathscr{F}$ (closure under countable unions).

The probability measure P is a set function from $\mathscr{F}$ to the interval $[0, 1]$, $P : \mathscr{F} \to [0, 1]$, and it satisfies the following four axioms.

(1) *Unit normalization* $P(\Omega) = 1$.

(2) *Nonnegativity* $(\forall F \in \mathscr{F}) \, P(F) \geqslant 0$.

(3) *Finite additivity* $F_1, ..., F_n \in \mathscr{F}$ and pairwise disjoint (no points in common) imply $P(\bigcup_{i=1}^{n} F_i) = \sum_{i=1}^{n} P(F_i)$.

(4) *Continuity* $(\forall i) \, F_i \supseteq F_{i+1}$ (containment), $\bigcap_{i=1}^{\infty} F_i = \varnothing$ (empty set) implies $\lim_{i \to \infty} P(F_i) = 0$.

Kolmogorov also proposed a definition of stochastic independence, relating the notion of axiomatic probability to informal or intuitive ideas of unrelated, unlinked, or independent events or experiments, and a definition of conditional probability, relating axiomatic probability to an informal idea of the effect of the knowledge of the outcome of an event on the uncertainties, or chances, of the outcomes of other events. These two definitions, while not generally considered part of the axioms of probability, are nevertheless fundamental for the under-standing, development, and application of axiomatic probability theory.

The $(\Omega, \mathscr{F}, P)$ model and associated definitions are deemed by many to be virtually universally applicable in those situations where we have to deal with chance or uncertainty. Our chief purpose in this chapter is to argue against this position by criticism of the universality and adequacy of the Kolmogorov setup for probability. Our analysis of $(\Omega, \mathscr{F}, P)$ first presents arguments that the Kolmogorov setup is unnecessarily restrictive and exclusive of legitimate models of random phenomena. The analysis of P is developed from two viewpoints. The first is the formal one of comparative probability as outlined in Chapter II. We find that the unit normalization, nonnegativity, and additivity are constraints imposed upon P which have little intuitive justification in terms of the relation of chance and uncertainty exposed by the axioms of comparative probability. The second viewpoint is that of the empirical meaningfulness of the probability scale as explicated through measurement theory. We find that the probability scale, as axiomatized by Kolmogorov, is occasionally empirically meaningless and always embodies an arbitrary choice or convention. While conven-tions can be harmless, there is the danger that the apparent specificity of the Kolmogorov setup may obscure the absence of a substantial grip on the structure of random phenomena.

The arguments for the inadequacy and incompleteness of the Kolmogorov theory center on the need for more specific choices of event field and class of probability assignments, the dependent position of conditional probability, the sufficiency of the characterization of stochastic independence, the absence of a distinction between scales of chance and uncertainty and those of mass, length, etc., and the supposedly close relation to the relative-frequency interpretation of probability.

If we appear to be playing the "devil's advocate" in the ensuing discussion, it is justified by the scarcity of such arguments in the enormous engineering, statistical, and philosophical literature on probability.

IIIB. Overspecification in the Kolmogorov Setup: Sample Space and Event Field

1. The Sample Space Ω

Kolmogorov's formulation presumes that the "experimenter" can uniquely (without duplication) and exhaustively (without omission) index or list all of the outcomes of his "experiment" at a level of detail sufficient for his interests. It is also presumed that the experimental outcomes can be determined with "certainty." The ability to index outcomes uniquely may be a nontrivial assumption (e.g., it may require a nontrivial theorem to verify that two labels index the same object). An exhaustive list of outcomes may be difficult to obtain in a high uncertainty situation. For example, in coin tossing, should we include the possibility that the coin lands on its edge, is lost, etc. ? de Jouvenal [2] has commented on the difficulties encountered in the construction of a complete Ω for social and economic phenomena, a domain of increasing interest to engineers.

It might be thought that one could avoid this enumeration difficulty by simply listing several possibilities of particular interest $\omega_1, ..., \omega_N$ and then defining the outcome ω_{N+1} as any outcome other than $\bigcup_i \omega_i$; in this way one achieves logical completeness of $\Omega = \{\omega_1, ..., \omega_{N+1}\}$. However, it would generally be difficult to assign a probability to an elementary event with the sweeping implications of ω_{N+1} ; we may be willing to hazard the guess that in a single toss of a coin a head is as likely to result as a tail, but what do we know about the relative probability of losing the coin before completing the toss ? Furthermore, in the absence of a well-defined universal set, the introduction of ω_{N+1} is suggestive of Russellian set paradoxes. For these reasons we do not usually complete Ω in this manner. It is far more common to restrict Ω to a list of "interesting" outcomes concerning which we have sufficient data; for example, in coin tossing, the outcomes are usually only a head or a tail. A consequence of Ω being logically incomplete appears below in our discussion of the field of events.

The question of the degree of specificity of the elementary event ω is related to our supposed ability to discern with certainty which event occurred at the conclusion of the experiment. In any application of this formal model to practice we are required to judge what can be observed with certainty and to use only such data as the substrate for our Ω. Quantum mechanics provides a well-known instance in which the classical notions of what could constitute an ω (e.g., simultaneous assessment of position and velocity) were ruled out by the uncertainty principle. Presumably the selection of ω is determined by some concept of "practical certainty" corresponding to an unexplicated notion of uncertainty that is distinct from that notion of randomness subsequently captured by assigning P to the elements of $\mathscr{F}$. There is something unsatisfactory about founding a theory of uncertainty upon a different, and unexplained, notion of uncertainty. The working consequence of the need for "practical certainty" is that too high a degree of conservatism in this regard can reduce our ability to describe outcomes to but the very coarsest and uninteresting categorizations. Possibly a more widely useful setup would have axioms which were reasonable even when ω could not be determined exactly at the conclusion of an experiment. See Savage [3] for further comment on the selection of a sample space.

2. The σ-Field of Events $\mathscr{F}$

The field of events is the collection of "macroscopic" events or the subsets of Ω whose occurrence is of interest to us and about which we have information. The whole set Ω is taken as the event corresponding to the performance of the experiment without regard to the particular outcome. While a field need not contain more than Ω and the null (empty) set $\varnothing$ (indicating the nonperformance of the experiment), it is usual that there are other events of interest. If we are interested in an event F, need we also be interested in its complement with respect to Ω, $\bar{F}$? At this point let us again distinguish between the complement $\bar{F}$ of an event F and the negation or nonoccurrence of the event. These two ideas are only equivalent when, as is not common in practice, Ω is a list of all (logically) possible outcomes. If Ω contains only those outcomes (elementary events) of reasonable interest, then information about the occurrence of F does not simply carry over to $\bar{F}$; $\bar{F}$ is no longer the nonoccurrence of F but rather the occurrence of a particular set of "interesting" outcomes in Ω. Inclusion of the complements of events in $\mathscr{F}$ would be more strongly indicated in a theory of the logical type, to be discussed subsequently, than in an axiomatic basis for an empirical theory; while a logical theory might well require closure under negation, an empirical theory of experiments need not necessarily concern itself with the nonoccurrence of all possible events.

The question of closure under unions or intersections is a more substantial issue. If F_1 and F_2 are in $\mathscr{F}$, then so are $F_1 \cup F_2$ and $F_1 \cap F_2$. Yet there occur circumstances in which we can reasonably claim to know $P(F_1)$ and $P(F_2)$ but know neither $P(F_1 \cup F_2)$ nor $P(F_1 \cap F_2)$. A class of photographs may be such that the probability of a predominantly dark picture is $\frac{1}{2}$ and the probability of a grainy texture is also $\frac{1}{2}$. However, we may have no data on, and little interest in, whether grainy patterns tend to be dark or not. Should we then be prevented from using probability theory to model this sample source when designing pattern classifiers for this problem?

At this point the initiate may interject that we can "parametrize" the family of pattern class distributions to account for our ignorance of certain probabilities. Nonetheless, why are we better off proceeding as if we have the required information and then introducing more or less *ad hoc* principles of statistics to account for our lack of information than by applying a theory which from the start accepts the possibility that we may lack certain information and be uninterested in certain combinations of events?

The central reason for the inappropriateness of the concept of a field for the collection of random events is that it emphasizes the aspect of the collection as containing all of the interesting events to the disadvantage of the requirement that we have data on the occurrences of the events in the field and can assign probabilities to them. If we think of the random events as subsets of Ω described intensively [i.e., A is the set of elements for which a given proposition p is true, $A = \{\omega : p(\omega)\}$], then the suitability of a field follows from accepting the proposition "not-p" as interesting once we accept "p" as interesting and from accepting the propositions "p or q" and "p and q" as interesting once "p" and "q" are individually accepted as interesting. However, by so doing we may include sets as events for which we are unable to assign probabilities, and we thereby conflict with an important limitation on a suitable collection of random events. The notion of a Λ-field, to be introduced shortly, avoids this problem by requiring the introduction of only those new events whose probability can be calculated from the given probabilities through the Kolmogorov axioms.

A further step in the specification of $\mathscr{F}$ is the requirement of countable unions of sets in $\mathscr{F}$ being in $\mathscr{F}$. Requirements involving limits are not derived from a desire to model chance phenomena accurately but rather from the grounds of "mathematical convenience." To quote Kolmogorov[†] [1, p. 18], "Thus sets of $B\mathscr{F}$ [the σ-field] are generally merely ideal events to which nothing corresponds in the outside world." He then points out that such an idealization should not yield contradictions with practice. However, this gain in analytical convenience (essentially closure under monotone limits) is at the price of having to deal with a collection of events that may, as pointed out by von Mises, contain some events whose outcomes are physically unverifiable and omit other events whose probability can be experimentally ascertained to any desired degree of precision.

3. The Λ- and π-Fields of Events

Three possible alternatives to $\mathscr{F}$ are the Λ-field and π-field described by Dynkin [4] and the von Mises field, $\mathscr{V}$ (denoted F_1 by von Mises). The Λ-field is defined as follows:

(1) $\Omega \in \Lambda$;

(2) $F \in \Lambda \Rightarrow \bar{F} \in \Lambda$;

(3) If $F_i \in \Lambda$ and $\{F_i\}$ are pairwise disjoint, then $\bigcup_i F_i \in \Lambda$.

A simple example of a Λ-field for $\Omega = \{a, b, c, d\}$ is given by $\Lambda = \{\varnothing, \Omega, (a, b), (c, d), (a, d), (b, c)\}$ where $\varnothing$ is the null or empty set. We may interpret this example in terms of the photograph pattern classification problem mentioned earlier as follows:

a predominantly dark color and grainy texture;
b predominantly dark color and smooth texture;
c predominantly light color and smooth texture;
d predominantly light color and grainy texture.

This Λ-field of events includes only those events corresponding to the pattern classes of dark, light, smooth, or grainy photographs, and it does not include such, more complex, patterns as a dark and grainy photograph. In the usual event field $\mathscr{F}$ we would have to include the event of a dark and grainy photograph as soon as we admitted (a, b) and (a, d) as events.

It is evident that the weaker assumption of a Λ-field avoids having

---

† A. Kolmogorov, *Foundations of the Theory of Probability*, p. 18. Bronx, New York: Chelsea, 1956.

to complete a probability assignment unnecessarily, and yet it represents a collection of events for which the probability axioms (particularly countable additivity) are well defined and reasonable. In fact the axioms for Λ require us, given events $\{F_i \in \Lambda\}$, to include only those other events whose probabilities are determinable solely from the given $\{P(F_i)\}$ and the Kolmogorov axioms for probability. Furthermore, from a mathematical viewpoint, Λ-fields are monotone classes (monotone sequences of sets which are ordered by inclusion have limits in the class) and would allow discussion of limits. However, Λ is not particularly suited for the definition of conditional probability. It is possible that $F_1, F_2 \in \Lambda$ but $F_1 \cap F_2 \notin \Lambda$; thus we could not define $P(F_2 \mid F_1)$.

The π-field is defined by

$$F, G \in \pi \Rightarrow F \cap G \in \pi,$$

or if we wish to ensure closure under monotone decreasing limits,

$$(\forall i) F_i \in \pi \Rightarrow \bigcap_{i=1}^{\infty} F_j \in \pi.$$

The advantage of the π-field over the Λ-field is that it enables us to define conditional probabilities. A disadvantage of the π-field is that we may have to include sets as events whose probabilities are not readily available.

4. The von Mises Field of Events

The von Mises field $\mathscr{V}$ arises from a careful study of experiments consisting of repeated, unrelated trials, each trial having a finite number of possible outcomes (e.g., repeated coin tossing). One may assume that such experiments and the relative-frequencies of occurrence of the various outcomes lie at the bottom of the Kolmogorov system [1, p. 4]. Hence it is significant that von Mises' investigations [5] led him to value a field of events $\mathscr{V}$ which, in general, differs from a σ-field. The sample space for the outcomes of indefinitely repeated experiments, each experiment having outcomes in Ω, is taken as the set Ω^{∞} of all infinite sequences with elements $\omega_i \in \Omega$. All events whose occurrences are determined by the first n experiments (cylinder sets) are in the field $\mathscr{V}$. We then construct the σ-field $\mathscr{F}_n$ of unions and complements of events depending on the first $m \leqslant n$ outcomes, and define $\mathscr{F}' = \bigcup_{n=1}^{\infty} \mathscr{F}_n$; $\mathscr{F}'$ is a field (no longer a σ-field) all of whose elements are finite coordinate cylinder sets. The field $\mathscr{F}'$ can then be extended to $\mathscr{V}$ as follows: $\mathscr{V}$ is the smallest field satisfying

$$(\forall n)(\exists L_n \,,\, U_n \in \mathscr{F}_n)\,(L_n \subseteq V \subseteq U_n)(\lim_{n\to\infty}[P(U_n) - P(L_n)] = 0) \Rightarrow V \in \mathscr{V}.$$

The field $\mathscr{V}$ contains sets measurable in the sense of Peano and Jordan. The events in $\mathscr{V}$ have the important property that their probability can be "measured" to any desired degree of accuracy by repetitions of verifiable events (events whose outcomes are decided by a finite number of experiments). Events not in $\mathscr{V}$ do not have experimentally assessable (by relative-frequencies) probabilities.

If, for example, we take the coin-tossing experiment, then the event $V = \{$heads occur only finitely often$\}$ is not in the von Mises field although it is in a σ-field generated from the cylinder sets. The tail event V does not depend on the first n outcomes of the repeated experiment for any finite n. Thus L_n must be the null set and U_n the sample space and $\lim_{n\to\infty}[P(U_n) - P(L_n)] = 1$. Hence the von Mises field $\mathscr{V}$ does not include the minimal σ-field generated from the cylinder sets. It is also true that the minimal σ-field need not include $\mathscr{V}$, and $\mathscr{V}$ is contained in the completion of the minimal σ-field. Von Mises' analysis successfully challenges the axiomatic imposition that the proper collection of events must be a σ-field.

IIIC. Overspecification in the Probability Axioms: View from Comparative Probability

One approach to an understanding of the restrictive and arbitrary aspects of the Kolmogorov axioms for quantitative probability is to start from the weaker and, therefore, less controversial concept of comparative probability and build up to the Kolmogorov concept of probability.

1. Unit Normalization and Nonnegativity Axioms

As can be seen from axiom C5 and Theorem II.1, it is an assumption that uncertainty or chance can be quantitatively described, although one we accept. Even when the CP relation $\gtrsim$ satisfies C1–C7, the agreeing quantitative probability P is not uniquely determined. Any strictly increasing transformation f of P, where $P' = f(P)$, results in an agreeing probability P'. This lack of unicity allows us to choose P such that

$P(\Omega) = 1$, $P(A) \geqslant 0$, thereby satisfying the first two Kolmogorov axioms. However, nothing necessitates this choice; it is simply a convention. The problem with conventions is that we may lose sight of their arbitrary origin. The unit normalization and nonnegativity conventions begin to appear as substantive properties and much statistical and computational effort is expended to ensure that these properties hold for estimates of, and calculations with, probability.

2. Finite Additivity Axiom

As we have commented in Section IIC, we are not convinced that orderings of random events must necessarily admit of finitely additive representations. The assumption that P must be additive is restrictive of the notion of CP; for example, we must rule out cases such as the Kraft *et al.* example (∗). The discussion of Scott's theorem in Section IIC and of the empirical meaning of additivity to be given in Section IIID suggests that the property of additivity is neither logically necessary nor necessarily meaningful.

The rationale underlying the almost universal choice of a finitely additive probability seems to be compounded from considerations of the important relative-frequency interpretation and the intuitive but incorrect conclusion that we require additivity to reflect the properties asserted in the comparative probability axioms C1–C4. As we have previously noted, a relative-frequency-based measure of chance is compatible with a CP relation $\gtrsim$ that admits of no additive agreeing probability; for example, if N_A is the number of occurrences of event A in unlinked repetitions of an experiment, then there are CP relations satisfying

$$N_A > N_B \Rightarrow A \gtrsim B$$

for which no agreeing P is additive. A more general approach to the resolution of the question of the necessity for a CP relation to have an additive agreeing probability is to adopt a rational method for the estimation of CP and then to see if it inexorably leads to additivity. Reflection on the decision theory sketched in Section IIG and its extensions suggests that the particular CP relation that best fits our prior knowledge about some random phenomenon and our available observations will not necessarily be compatible with additivity. A blind insistence upon modeling chance phenomena only through additive probability forces us to ignore potentially preferable descriptions.

Finally, even if we agree to restrict CP to only those relations admitting of additive P, the choice of additivity as distinct, say, from multiplicativity is another arbitrary convention. As discussed in Subsection IIID3, there are infinitely many probability scales equivalent to the additive scale and no significant conceptual basis for preferring one scale to the other.

3. Continuity Axiom

If $\gtrsim$ satisfies C8, then any finitely additive P will be continuous and, by a well-known theorem in probability, also be countably additive. While C8 is not unreasonable, this technical requirement has an appreciable effect in ruling out possible probability assignments. As is well known, it is not always possible to assign arbitrary countably additive probabilities to arbitrary collections of subsets of Ω nor even to arbitrary σ-fields. By the Caratheodory extension theorem [6] we can always uniquely extend a countably additive probability specified over a field to the smallest σ-field containing the field and hence to the completion of the σ-field. In general we cannot extend the assignment of P to other sets, although specific assignments can often be extended somewhat. For example, there is no countably additive measure extendible to all subsets of Ω and assigning intervals in $\Omega = (0, 1)$ their length.

A more interesting example is provided by the experiment of "drawing a positive integer at random." By drawing a positive integer at random we might mean in part that $\Omega = (1, 2, 3, ...)$, $\mathscr{F}$ is the power set of Ω (set of all subsets of Ω), and P assigns zero probability to all finite subsets of $\mathscr{F}$. If P is countably additive, then, by the Caratheodory extension theorem, P extends to $\mathscr{F}$ by being identically zero, in violation of $P(\Omega) = 1$. Thus there is no countably additive measure capable of describing this somewhat idealized experiment. One resolution of this difficulty is provided by Renyi's axiomatization of conditional probability described below. A second resolution is to relax the unit normalization and allow, say, σ-finite measures, although this could violate C4 and C6. A third resolution is to replace countable additivity by finite additivity. If we have only finite additivity, we cannot conclude from $P(\omega_i) = 0$ that $P(\Omega) = 0$. We might define P for those subsets F of Ω for which the following limit (density) exists: Let $N_F(n)$ be the number of elements of F that are less than or equal to n and define $P(F) = \lim_{n \to \infty} (1/n) N_F(n)$. This yields a nonnegative, finitely additive measure such that $0 \leqslant P(F) \leqslant P(\Omega) = 1$.

Nor can we find a justification for countable additivity in an unsophisticated appeal to the limit of a relative-frequency interpretation of probability. If $\Omega = \{\omega_i\}$ and $N_{\omega_i}(n)$ is the number of occurrences of ω_i in the first n trials, then it does not follow that

$$\sum_i \lim_{n \to \infty} \frac{N_{\omega_i}(n)}{n} = \lim_{n \to \infty} \sum_i \frac{N_{\omega_i}(n)}{n}.$$

In fact, while

$$\sum_i \frac{N_{\omega_i}(n)}{n} = 1,$$

it is possible for

$$(\forall i)\, P(\omega_i) = \lim_{n \to \infty} \frac{N_{\omega_i}(n)}{n} = 0.$$

IIID. Overspecification in the Probability Axioms: View from Measurement Theory

1. Fundamentals of Measurement Theory

The examination of the relation between comparative and quantitative probability exposes some of the formal conditions that make it possible to choose P satisfying the Kolmogorov axioms. It does not, however, show why that choice is either meaningful or necessary. Choosing the Kolmogorov axioms for P amounts to, at the least, a choice of a measurement scale for the random phenomena of chance and uncertainty. To better appreciate what is involved in choosing a meaningful measurement scale, we digress and follow Suppes' and Zinnes' treatment of measurement theory [7], as also treated in [8].

Fundamental measurements can be thought of as establishing a homomorphism between a given empirical relational system $\mathscr{E}$ and a selected numerical relational system $\mathscr{N}$.

The $\mathscr{E}$ system consists of a domain of real objects or events together with a collection of (significant) relations between elements of the domain; for example, the domain might contain many masses and a relation might be that of "heavier than" as determined by a balance scale. The $\mathscr{N}$ system consists of a domain of real numbers together with selected numerical relations; for example, the domain might be the positive reals and a relation might be that of "greater than." The measurement system is a triple $(\mathscr{E}, \mathscr{N}, f)$ with a mapping f (homomorphism) between the domains of $\mathscr{E}$ and $\mathscr{N}$ that preserves corresponding empirical and numerical relations; for example, by means of a spring scale we can assign positive numbers to the masses such that "heavier than" between masses corresponds to "greater than" between their numerical weights.

Measurement systems can be classified according to the groups of transformations of f that leave unchanged the correspondences between relations. For example, any strictly increasing function of f would preserve a correspondence between "heavier than" and "greater than"; such a measurement system is called an *ordinal* scale. An *absolute* scale is one for which only the identity transformation preserves the relational homomorphisms. There are few physical concepts that require an absolute scale. The number of objects in a collection requires an absolute scale, but mass, length, time, temperature, current, voltage, etc., do not.

To clarify the measurement status of probability let us first discuss the familiar concept of weight. We assume that we have a set of individual masses $a, b, c, \ldots$. There is an operation of combining or packaging masses together. By an object we mean either an individual mass or a package of masses from our collection. The basic empirical relation between masses is established through a two-pan balance scale. The measurement of weight, ω, as carried out perhaps by a spring scale, assigns positive numbers to objects. If we are only interested in whether an object A is heavier than an object B, then the measurement scale ω need only be such that "A is at least as heavy as B," if and only if, "$\omega(A) \geqslant \omega(B)$." In such an ordinal scale the assertion "$\omega(A)/\omega(B) = k$" has no meaning beyond that of the implicit inequality. For the numerical statement "$\omega(A)/\omega(B) = k$" to be empirically meaningful there must be an empirical relation that is transformed by the function (scale) ω into it.

In order to interpret the numerical ratio statement we might introduce an empirical concept of a standard series of weights modeled on Luce's definition of a standard series given in Section IIB. By a standard series of weights $\{B_i\}$ from object B to object A we mean the following: B_1 balances B; C_j balances B; C_j and B_j contain no common object; B_{j+1} balances the package of C_j and B_j ; and for some n, B_n balances A. (Note that we permit an object to "balance" itself.) We can now supply an empirical interpretation for "$\omega(A)/\omega(B) = k$" as follows: There is an object C such that there is a standard series of length n from C to B and of length kn from C to A.

We can also interpret the numerical difference statement "$\omega(A) - \omega(B) = z$" as corresponding to the following empirical relation: There are objects C, D, and E, where C balances B, C and D

have no objects in common, E balances A, the package of C and D balances E, and $\omega(D) = z$.

The preceding constructions do not exhaust the possibilities for establishing an empirically meaningful additive scale. More generally we might conclude that ω is an empirically meaningful additive scale if, to within transformation through multiplication by a positive constant, it is the unique scale satisfying

(a) A is at least as heavy as $B \Leftrightarrow \omega(A) \geqslant \omega(B)$,

(b) $A \cap B = \varnothing \Rightarrow \omega(A \cup B) = \omega(A) + \omega(B)$.

The fundamental ideas of measurement scales are presented in some detail in Krantz *et al.* [8]. If our criterion of empirical meaningfulness is violated, then statements of ratios or of differences are happenstance and need have no empirical significance beyond that of establishing inequality.

2. Probability Measurement Scale

To apply the preceding discussion of weighing directly to probability we correspond elementary outcomes to individual masses, Ω to the collection of masses, events to objects, and the operation set union to packaging masses. The probabilistic analog to the balance scale which will enable us to compare empirically the probabilities of events depends on our interpretation of probability (e.g., relative-frequency of occurrence, willingness to wager on occurrence, etc.). Without involving ourselves at this moment in interpretive questions, let us assume the existence of some mechanism that can determine the CP relation $\succsim$. Probability is then a mapping in a measurement system where the domain of $\mathscr{E}$ is a field of events and the domain of $\mathscr{N}$ is the unit interval $[0, 1]$. The empirical relations concern the uncertainty, or chance, of the occurrence of an event in the performance of an experiment.

Reasonable empirical interpretations for ratio or difference statements of probabilities follow directly from our discussion of weight. However, if we are unable to find suitable standard series, etc., then the usual probability statements are meaningless numerical accidents. Now, while it is common in thinking of weighing to imagine that we can create a large collection of weights and thereby find the needed standard series, etc., this is not the case in applications of probability. In applications of probability theory we generally deal with a given possibly finite set Ω and field $\mathscr{F}$. It is evident that for finite Ω we cannot generally expect that ratio and difference statements will be meaningful. For instance, if $\Omega = \{0, 1\}$ with $\varnothing \prec 0 \prec 1 \prec \Omega$, then there is no empirical meaning in $P(1) = 3P(0)$, beyond the information that $P(1) > P(0)$. In fact, it is only when Luce's axiom (L) for CP (Section IIC) is satisfied that an empirical interpretation of ratio and difference statements can be given using standard series. Nor is there any way of satisfying our general criterion for the empirical meaningfulness of an additive scale, except in the case of $P(1) = P(0)$. How can this conclusion be reconciled with the widespread belief in the meaningfulness of Kolmogorov's probability function for finite probability models?

One possibility at reconciliation is that there may be another empirical relation that maps into, and thereby interprets, the usual numerical probability statements. Concerning this possibility we have no considered ideas.

Another possibility is that finite probability models may, in reality, be elliptical references to models with larger or even infinite sample spaces. For example, we may assert that a random experiment is such that $\Omega = \{0, 1\}$, $P(1) = .75$. A relative-frequentist would claim that $P(1)$ was derived by examining the more complex random experiment having as realizations n-tuples of binary outcomes describing independent repetitions of the original experiment. Hence the probability statement originates with an experiment having a sample space with 2^n points, rather than 2 points, in which it is meaningful.

The crux of this explanation can be put in a form that does not depend on a particular interpretation of the probability concept. Savage and Kraft *et al.* in their polarizability axiom (see Section IIC) have considered the possibility of establishing the existence of an additive P representation for a random experiment by requiring that the CP relation $\succsim$ be extendible in a consistent manner to the compound experiment that augments the given experiment with a coin-tossing experiment. This device can also be used to interpret the additive P scale. Let $\Omega = \{\omega\}$ and $\succsim$ describe a random experiment. Let $\Omega_n = \{\omega_n\}$ be a sample space for n "independent" tosses of a "fair" coin; that is, in whatever interpretation we have for probability, the tosses are such that all outcomes are equally probable, $\omega_n \approx \omega_n'$. The compound experiment has a sample space $\Omega^* = \{(\omega, \omega_n)\}$ and an ordering $\succsim^*$. The ordering $\succsim^*$ is consistent with $\succsim$ in that

$$(\omega, \omega_n) \succsim^* (\omega', \omega_n') \Leftrightarrow \omega \succsim \omega'.$$

It may now be possible to make meaningful numerical statements about the compound experiment that were not possible in the original experiment. For example, we can meaningfully claim that $P(1) = 3P(0)$, or $P(1) = .75$, when $\Omega = \{0, 1\}$ by referring it to the empirical proposition

$$(1, HH) \approx^* (0, TT) \cup (0, TH) \cup (0, HT).$$

The explanation above is not an argument for the general acceptability of the polarizability axiom or for a freedom to always associate an independent, fair coin-tossing experiment with any given random experiment. This would put the cart before the horse. We have argued that the usual finite probability model $(\Omega, \mathscr{F}, P)$ may suppress the description of the more complex finite, or infinite, probability model that provided the information on P. The introduction of the coin-tossing experiment is just a formal device to generate a canonical reconstruction of the complex, original random experiment. In the more realistic finite Ω models of random experiments, using CP in place of quantitative probability, there may never have been a more complex "parental" experiment. In this case the canonical reconstruction, even when possible, is not compelling in its conclusions; we cannot create meaning out of nothingness.[†]

Unlike the case of mass, length, and so forth, we do not believe that the tendency for an event to occur can always be split into equal pieces.

It appears that the Kolmogorov probability scale may be unnatural when we deal with random experiments having finitely many outcomes. Not only may an additive scale not exist (e.g., the counterexample of Kraft *et al.* in Section IIC), but even when it exists it may be empirically meaningless. The inadequacy of the Kolmogorov axioms, from this point of view, appears less acute when we deal with "reasonable" infinite sample space models.

3. Necessity for an Additive Probability Scale

Even if we are able to interpret the Kolmogorov absolute probability scale meaningfully we have yet to argue for the necessity of this scale. In fact, there are many equivalent scales that are not additive. From

---

[†] An analogy may clarify this point. The reader, by invoking his knowledge of vocabulary and grammar, understands the statement, "It's a beautiful day, and I feel fine." However, if he now learns that this statement appears in a long list generated by a randomly programmed computer, would he still think it appropriate to evaluate the meaning of the statement by bringing the same language tools to bear? In the view we have been proposing, the statement has meaning in the usual case because we assume that it was made by a person. Learning that a computer generated it does not change the statement but radically changes its meaning.

Theorem II.4, we know that, under reasonable axioms for comparative probability, there exist P, G, and

$$A \cap B = \varnothing \Rightarrow P(A \cup B) = G(P(A), P(B)).$$

It can be shown [9] that under mild regularity conditions the strictly increasing, symmetric, associative function G can be represented in the quasilinear form

$$G(x, y) = h(h^{-1}(x) + h^{-1}(y)),$$

where h is some strictly increasing function. From the quantitative probability P we can derive an additive probability P' through the definition $P'(F) = h^{-1}(P(F))$. The function G' corresponding to P' is then $G'(x, y) = x + y$. What real considerations though make us prefer the particular choice of measurement scale h^{-1} yielding an additive, quantitative probability P'?

If we think of probability as a model of relative-frequency, then we seem to be led to accept the choice of $G(x, y) = x + y$. However, agreement on probability as built upon relative-frequency as the quantitative core of chance phenomena does not eliminate the possibility that P may depend on some function of the relative-frequency other than the identity function. For example, we might interpret $P(F)$ as $\lim_{n \to \infty}(N_F(n)/n)^2$, where $N_F(n)$ is the number of occurrences of the event F in n trials. In this case we would find that F_1 and F_2 disjoint imply that $P(F_1 \cup F_2) = [\sqrt{P(F_1)} + \sqrt{P(F_2)}]^2$, the scale transformation h^{-1} taking P into an additive probability assignment being $h^{-1}(x) = \sqrt{x}$. If we accept the fact that there is exactly as much information about a chance event F in $(N_F(n)/n)^2$ as in $N_F(n)/n$, then it becomes only a matter of habit or convenience as to which of these we prefer to work with—a matter of choice and not necessity. (But see Section IVJ.)

An analogy to this situation is that of coordinate-free results in analytic and vector geometry. The choice of an additive scale is analogous to choosing a fixed coordinate system and expressing all results in that system, although the results do not depend on the particular system chosen. The danger in this procedure when applied to probability is the ease with which we forget that the results are coordinate-free and then attempt to make something of the choice of system. The Kolmogorov scale for probability is in large part a convention and occasionally its use evokes empirically meaningless statements.

In sum the observations of Sections IIIC and IIID suggest that the Kolmogorov axioms for probability rule out of consideration possible orderings of random events and then adopt conventions with regard to the remaining orderings of events. The conventions make the choice of probability scale spuriously specific and make attempts to extract significant probabilistic relations from data more difficult. The user finds himself forced to engage in arbitrary assumptions, to match those of the theory, which appear naked in the light of a real application.

IIIE. Further Specification of the Event Field and Probability Measure

1. Preface

Thus far our remarks have indicated that the widely used Kolmogorov setup may be unnecessarily restrictive in certain applications. However, it can also be argued that this theory needs additional axioms before it can be reasonably and profitably employed to model random phenomena; the axioms say little and some of what they do say is either unimportant or a matter of convention. There are several points at which the Kolmogorov axiomatic theory needs to be augmented and discussed.

(1) In applications one generally chooses a collection of events and a family of possible probability assignments, prior to any analysis. What are the guidelines for these choices?

(2) Should the concepts of stochastic independence and conditional probability enter probability theory merely as definitions or do they warrant a presentation coordinate with, or even prior to, that of absolute probability?

(3) What distinguishes probability and its domain from, say, length or mass?

We only hope to clarify, and not answer, these questions in the remainder of this chapter. The lack of an adequate set of answers will partially motivate us to consider, in subsequent chapters, other approaches to probability.

2. Selecting the Event Field

Guidelines for the selection of an event field have been offered in the important cases of experiments with outcomes describable by either finitely, countably infinitely, or uncountably many random variables. The σ-field usually assumed when we deal with the collection of events for a real random variable is the Borel field of subsets of the real line; Cramer [10] has gone so far as to include this suggestion in the axioms for probability. The Borel field is the smallest, in the sense of inclusion, σ-field containing all intervals; the field may then be completed for a given probability measure by including sets that differ from sets in the Borel field by sets of zero probability. The Borel field is generally thought to be sufficiently rich in events for most problems. Furthermore, it is felt that since the Borel family of events is generated from the simple and basic events of the experimental outcomes being greater than or less than a given number, they are particularly appropriate for the description of experiments.

The Borel field is also the usual choice when dealing with sequences of random variables, for example, repeated experiments with real-valued outcomes. The Borel field of events for an infinite sequence of random variables is the smallest σ-field containing, for every finite n, all of the cylinder sets with n-dimensional intervals as bases. In modeling random processes we encounter the most general case of collections of random variables. It has been suggested that in such instances the Borel field of events be restricted to be standard, that is, the event field should be point isomorphic to some Borel field of subsets of the line [11]. Whatever the adequacy and naturalness of the Borel field for finitely many random variables, von Mises would challenge its use for an infinite sequence of random variables.

An important restriction on the von Mises field $\mathscr{V}$, absent from the definition of σ-field, is that of "conceptual verifiability." An event for an experiment with indefinitely repeated trials is considered to be conceptually verifiable if it is either determined by a finite number of trials or can be approximated by events whose outcomes are determined by a finite number of trials. Von Mises restricts the class of events to those whose probability can conceivably be estimated from the relative-frequency of their occurrence. If an event cannot be approximated by events whose outcome is determined in a finite time, then it is unreasonable to think of repetitions of this event. While von Mises' position is a reflection of his study of single sequences of repeated trials, recourse to an ensemble of repeated experiments does not remove the difficulty. It will still take an infinite time before we gain even an approximate idea of the relative-frequency of occurrence of an event not in $\mathscr{V}$.

3. Selecting P

The desire for a more structured theory of probability is manifest in a suggestion of Gnedenko and Kolmogorov [12] for the modeling of a nondenumerable collection of random variables (random process).

If we consider the selection of a generally suitable event collection for a random process, then we have the complicated situation in which Ω is an infinite-dimensional function space. Gnedenko and Kolmogorov [12, pp. 18–19] suggest that in addition to the usual axioms one should add the requirements that for a topological space Ω, all the open sets of Ω (the topology $\mathscr{F}$) are in the σ-field $\mathscr{F}$, and $F \in \mathscr{F}$ implies $P(F) = \inf_{F \subseteq T \in \mathscr{F}} P(T)$. This idea is then extended to arbitrary abstract sets Ω that are not topological spaces by requiring that for all $x : \Omega \to R^1$ measurable, the induced measures in R^1 on the Borel field of sets satisfy the last stated, reasonable infimum property. The introduction of this requirement of what are called perfect measures reduces the apparent scope of probability theory in exchange for a more reasonable probability assignment.

An example given by Doob [12, pp. 245–251] shows that defining two random variables x_1 and x_2 as independent if and only if $P(x_1 \in F_1, x_2 \in F_2) = P(x_1 \in F_1) P(x_2 \in F_2)$ for all $F_1, F_2 \in \mathscr{F}$ is equivalent, when P is a perfect measure, to defining them as independent if and only if the distribution factors as a product for F_1 and F_2 intervals, but not necessarily equivalent otherwise; they are always equivalent, however, when $\mathscr{F}$ is a Borel field. Restricting $\{F_i\}$ to be intervals is the usually stated condition that the joint cumulative distribution function factors as a product of the marginal distribution functions. Other remarks of a technical character on the scope of the usual probability measure are presented by Doob [13] (e.g., separability) and Ito [11].

IIIF. Conditional Probability

1. Structure of Conditional Probability

Conditional probability enters the Kolmogorov theory as a definition in terms of absolute probability. However, one can argue that conditional probability is at least as fundamental as absolute probability and deserving of an independent development. The observation that all empirical absolute probabilities are conditional upon data suggests the importance of conditional probability. Of course, strictly speaking, the data need not be an event, and this is only an analogy. If we directly axiomatize conditional probability, then we arrive at a structure from which absolute probability is easily derived, and one which enables us to model a greater variety of random phenomena than permitted by absolute probability. Curiously the situation is rather different in comparative probability, as was discussed in Section IIE.

In Section IIE we presented axioms QCC0–QCC7 for a quaternary comparative conditional probability (QCCP) relation $\succsim^*$. We were then able to show through Theorem II.8 that

$$P(A \cap B \mid C) = G(P(B \mid C), P(A \mid B \cap C))$$

for G symmetric, associative, and increasing. However, while this result is similar to the product rule of the Kolmogorov theory, it is not reducible to it without other constraints on $\succsim^*$. These other constraints were not judged by us to have sufficient intuitive appeal to warrant inclusion in the theory of QCCP. Hence, we find the Kolmogorov definition of conditional probability both restrictive in its domain (not all $\succsim^*$ relations have an agreeing Kolmogorov conditional probability) and containing a large measure of convention (i.e., the choice of measurement scale).

Our examination of quaternary conditional comparative probability also revealed a less simple relationship to CP than one would expect from the usual quantitative probability definition. Not all CP relations $\succsim$ admit of a QCCP relation $\succsim^*$ related to it in the usual way that

$$A \succsim B \Leftrightarrow A/\Omega \succsim^* B/\Omega.$$

Furthermore even when a CP relation $\succsim$ is compatible with a QCCP relation $\succsim^*$, the latter need not be uniquely determined by the former. Seemingly neither CP nor QCCP is the more fundamental and both are required.

Renyi [14] has suggested a direct axiomatization of quantitative conditional probability. In brief, Renyi's conditional probability system $(\Omega, \mathscr{F}, \mathscr{G}, P(F \mid G))$ is defined as follows. The set Ω is the sample space of elementary events and $\mathscr{F}$ is a σ-field of subsets of Ω; $\mathscr{G}$ is a subset of $\mathscr{F}$ having the properties:

(a) $G_1 , G_2 \in \mathscr{G} \Rightarrow G_1 \cup G_2 \in \mathscr{G}$,

(b) $\exists \{G_n\}, \bigcup_{n=1}^{\infty} G_n = \Omega$,

(c) $\varnothing \notin \mathscr{G}$.

P is the conditional probability function satisfying the following four axioms.

R0. $P : \mathscr{F} \times \mathscr{G} \to [0, 1]$.

R1. $(\forall G \in \mathscr{G})\, P(G \mid G) = 1$.

R2. $(\forall G \in \mathscr{G}),\, P(\cdot \mid G)$ is a countably additive measure on $\mathscr{F}$.

R3. $(\forall G_1,\, G_2 \in \mathscr{G})\, G_1 \supseteq G_2 \Rightarrow P(G_2 \mid G_1) > 0,$

$$(\forall F \in \mathscr{F})\, P(F \mid G_2) = \frac{P(F \cap G_2 \mid G_1)}{P(G_2 \mid G_1)}.$$

We define absolute probabilities P^* satisfying a Kolmogorov system $(\Omega, \mathscr{F}, P^*)$ by selecting any $G \in \mathscr{G}$ ($\mathscr{G}$ nonempty by hypothesis) and setting $P^*(F) = P(F \mid G)$. If $\Omega \in \mathscr{G}$, then a more natural definition would be to use $P(F \mid \Omega)$. The advantage in selecting Ω is that for all $G \in \mathscr{G}$, $P(G \mid \Omega) > 0$, and it follows that $P(F \mid G) = P^*(F \cap G)/P^*(G)$ —the usual definition of conditional probability when we start from the Kolmogorov system.

As a simple application of the Renyi conditional probability system, we define the experiment, not describable by absolute probability, of selecting a positive integer at random. Let $\Omega = \{1, 2, 3, ...\}$, $\mathscr{F}$ be the set of all subsets of Ω, $\mathscr{G}$ be the set of all finite subsets of Ω, and $\|F\|$ be the number of elements in F (cardinality of F). Define $P(F \mid G) = \|F \cap G\|/\|G\|$. Thus the conditional probability space $(\Omega, \mathscr{F}, \mathscr{G}, P)$ defines the selection of a positive integer at random in a reasonable manner.

2. Motivations for the Product Rule

What motivates R3 or its parallel in the Kolmogorov system of the product rule for conditional probability? An answer to this question must be based on the roles and interpretation assigned to probability and conditional probability. One interpretation is that conditional probability describes a censored version of the random experiment described by probability. Assume that the events F and G are defined on the outcomes ω of a random experiment $\mathscr{E} = (\Omega, \mathscr{F}, P)$. Introduce the censored experiment $\mathscr{E}_G = (\Omega_G, \mathscr{F}_G, P_G)$ with outcomes ω_G given by

$$\omega_G = \begin{cases} \omega & \text{if} \quad \omega \in G \\ \in \varnothing & \text{if} \quad \omega \notin G; \end{cases}$$

that is, if $\mathscr{E}$ does not produce G, then the outcome is ignored, and it is considered that $\mathscr{E}$ was not performed. We define $\Omega_G = G \cap \Omega$, $\mathscr{F}_G = \{H : (\exists F \in \mathscr{F})\, H = F \cap G\}$. To relate P_G to P we assume that whatever interpretation is used to determine P for $\mathscr{E}$ will now be used to determine P_G for $\mathscr{E}_G$.

For example, if P is interpreted as a finite relative-frequency, then it is well known that $P_G(F) = P(F \cap G)/P(G)$. If P is interpreted as a limit of relative-frequency, then we cannot reach the same conclusion; the ordering of outcomes matters. If P is interpreted as a limit of relative-frequency for a repeated random experiment whose outcomes are a collective (see Section IVE), then it can be shown that $P_G(F) = P(F \cap G)/P(G)$. The subjective interpretation of P described in Section VIIIE also leads to the usual conclusion $P_G(F) = P(F \cap G)/P(G)$.

An interpretation of conditional probability through censored experiments corresponds to a view of conditional probability as a restriction of probability; $P_G(\cdot) = P(\cdot \mid G)$ is the renormalized restriction of P to the subfield $\mathscr{F}_G$ of $\mathscr{F}$. A different viewpoint, which is emphasized in theories of logical probability (Chapter VII), is that conditional probability is the formal mechanism by which we learn from experience. This is also the attitude of most workers on problems of pattern classification, adaptive, and learning systems. Cox [15] develops an expression for conditional probability $P(A \mid B)$ from the following axiom:

$$(\exists \text{ continuous } G)\, P(A \cap B \mid C) = G(P(B \mid C), P(A \mid B \cap C)).$$

It will be recalled that, except for the continuity of G, this axiom is a consequence of our axioms for QCCP, as shown by Theorem II.8. That is to say, the uncertainty in the occurrence of A and B given C is a function only of the uncertainties that B occurs given C and then that A occurs given the occurrence of B and C. Letting

$$x = P(B \mid A), \qquad y = P(C \mid A \cap B), \qquad z = P(D \mid A \cap B \cap C)$$

and examining alternative evaluations of $P(B \cap C \cap D \mid A)$ we arrive at the following identity:

$$G(x, G(y, z)) = G(G(x, y), z).$$

This functional equation can be recognized as the associativity equation mentioned in Section IIID. The solutions to this equation for G continuous x, y, z lying in a finite interval are of the form: $(\exists h$ continuous, strictly increasing$)$ $G(x, y) = h[h^{-1}(x) + h^{-1}(y)]$. The usual expression for conditional probability would follow from the choice $h(x) = \log x$. We are again faced with making what appears to be an arbitrary choice of scale. Exercise of this arbitrary choice is harmless only so long as we do not then impute a reasoned grounds for the specificity of the conclusion.

IIIG. Independence

1. Role of Independence

The idea of independence introduced through the definition of stochastic independence (SI) is one of the few ways in which the Kolmogorov system relates probability to chance. Kolmogorov's awareness of the role of independence is reflected in his comment [1, p. 9], "We thus see, in the concept of independence, at least the germ of the peculiar type of problem in probability theory." The informal idea of the independence of experiments or the irrelevance of the performance of one experiment for the outcomes of another experiment, and its formalization through the factorization of the probability measure for the joint experiment as a product of the probability measures for the component experiments, is frequently invoked in descriptions of chance phenomena. In practice we often have strong "feelings" that two experiments are unrelated, with neither influencing the other. The definition of SI enables us to use these beliefs, whatever their status, to reduce the family of possible probability distributions governing the joint experiment. A typical use of the assumption of SI is found in discussions of the laws of large numbers; the informal idea of independence is used to justify a detailed hypothesis concerning the appropriate probabilistic description for complex experiments involving indefinitely repeated trials.

The reasoning involved in the use of SI to determine probability has no counterpart in the determination of other quantities such as length or mass. The formal definition of independence is a substantive tie between chance and Kolmogorov's probability, and one which helps distinguish probability from other concepts whose measurement scales are describable in terms of measure theory. It seems indisputable that in engineering and elsewhere, independence plays a major role in the estimation of probabilities. However, we must be cognizant of the fact that invocations of SI are usually not founded upon empirical or objective knowledge of probabilities. Quite the contrary. Independence is adduced to permit us to simplify and reduce the family of possible probabilistic descriptions for a given experiment. The resulting probabilities then incorporate the *nonempirical* elements contained in the grounds we had for asserting SI. Assertions of SI arise from our understanding of random phenomena rather than from "empirical proof," although this is not to deny the frequent presence of explicit data suggesting independence. To quote Kolmogorov[1] [1, p. 9], "In consequence, one of the most important problems in the philosophy of the natural sciences is—in addition to the well-known one regarding the essence of the concept of probability itself—to make precise the premises which would make it possible to regard any given real events as independent."

2. Structure of Independence

Having remarked on one important role of the assumption of SI, we now consider the adequacy of the definition of SI for the concept of independence. In Section IIF we presented a formal structure for a concept of independence ($\perp$) related to comparative probability. We showed in Theorem II.10 and a subsequent lemma that if independence satisfied properties I1–I9, then A, B independent implied

$$P(A \cap B) = G(P(A), P(B)),$$

where G is symmetric, associative, and increasing in each variable. However, this conclusion was only a necessary condition for independence. In the absence of additional axioms for independence we can conclude neither that there is a P for which $G(x, y) = xy$ nor that the product factorization is a sufficient criterion to establish independence. Finally we commented that presently known axioms for CP and independence which guarantee the existence of an additive P and $G(x, y) = xy$ involve intuitively unacceptable constructions. We are unable to reach the Kolmogorov definition of independence starting from a more basic viewpoint. This suggests that there may be some arbitrariness in the Kolmogorov definition of the kind exposed in Sections IIIC and IIID for P.

A common argument for SI starts from the relative-frequency interpretation of probability. Let us assume that we have a joint experiment $\mathscr{E} = (\mathscr{E}_1, \mathscr{E}_2)$ and have observed "unrelated" repetitions of $\mathscr{E}$. Unrelatedness of the component experiments $\mathscr{E}_1$ and $\mathscr{E}_2$ is now taken to mean that for any possible outcomes A of $\mathscr{E}_1$ and B of $\mathscr{E}_2$, the occurrence of A is unaffected by the occurrence of B. A simple statistic with which to test for a change in the occurrence pattern of A induced B is by

[1] A. Kolmogorov, *Foundations of the Theory of Probability*, p. 9. Bronx, New York: Chelsea, 1956.

the number $N_{A \cap B}$ of simultaneous occurrences of A and B for a given number N_B of B. Unrelatedness of experiments implies that for any other outcome C of $\mathscr{E}_2$ such that $N_C = N_B$ we should find that $N_{A \cap C}$ is approximately equal to $N_{A \cap B}$; if this is not the case for all A, then knowledge of B for $\mathscr{E}_2$ tells us something about $\mathscr{E}_1$ other than what knowledge of C tells us. Thus we require for independence that the test statistic $N_{A \cap B}/N_B$ be approximately invariant with respect to the choice of outcome of $\mathscr{E}_2$. With the usual relative-frequency interpretation, we can restate our conclusion as $P(A \cap B)/P(B)$ is to be independent of B, or $P(A \cap B)/P(B) = f(A)$. Noting that $\bar{B}$ is an event when B is, and using the finite additivity property for P, yields $(\forall A \in \mathscr{F}_1, B \in \mathscr{F}_2) P(A \cap B) = P(A) P(B)$ as the well-known condition for $\mathscr{E}_1$ to be SI of $\mathscr{E}_2$. It is also apparent that SI is a symmetric relation.

The argument sketched above merely indicates that the usual definition of SI is a necessary condition for independence. Is it also a sufficient characterization of our intuitive understanding of independence? There is a remark in von Mises' discussion of randomness and independent trials to the effect that randomness is not fully prescribed by SI [5, p. 38]. However, it is unclear what else von Mises might have meant.

What may be the best viewpoint from which to discuss the meaning of the independence of experiments, or events, is provided by the computational-complexity approach presented in Section VH. If we consider two experiments $\mathscr{E}_1$ and $\mathscr{E}_2$ with finite sample spaces and unlinkedly repeat the pair of experiments n times, then we can record the outcomes of experiments $\mathscr{E}_1$ and $\mathscr{E}_2$ by sequences O_1 and O_2 of length n. If the experiments are independent, then knowledge of O_1 should be uninformative about O_2 and *vice versa*. The notion of "uninformativeness" is made precise through reference to a concept of computational complexity that in effect says we require about as large a program to compute O_2 on a suitable computer as we do when O_1 is given as supplementary data. This approach results in a new definition of SI that is strictly stronger than the usual definition. In particular, it appears that independence cannot be adequately characterized in terms of probability!

Finally, the definition of SI through the product factorization of the joint distribution is at best valid only for the usual choice of an additive probability scale. Had we elected to consider a probability assignment in some other scale h,

$$A \cap B = \varnothing \Rightarrow P(A \cup B) = h\big(h^{-1}(P(A)) + h^{-1}(P(B))\big),$$

then we would have had to modify the definition of independence in the following obvious manner: Events A and B are SI if and only if $P(A \cap B) = h(h^{-1}(P(A)) \cdot h^{-1}(P(B)))$.

IIIH. The Status of Axiomatic Probability

What are the contributions of axiomatic probability to an understanding of probability? What distinguishes characterizations of uncertainty and chance from characterizations of other quantities? Where do probabilities come from? In view of the manifold and extensive applications of axiomatic probability, it is remarkable how unsatisfactory the answers to these questions are. Stated baldly, axiomatic probability hardly enables us to understand what probability is about or what its subject matter is. The Kolmogorov axioms, for example, provide neither a guide to the domain of applicability of probability nor a procedure for estimating probabilities nor appreciable insight into the nature of random phenomena.

The apparent utility of the Kolmogorov theory is in fact due to its supplementation in practice by interpretive assumptions, many of which often go unstated. Some of these assumptions will be brought out into the open when we present the other approaches to probability. Others of the assumptions form the disciplines of statistics and the theory of rational behavior. The lack of essential content in, say, the Kolmogorov system imposes a great burden on these other disciplines. Within the discipline of probability itself the theorems of the laws of large numbers are often proposed as doing service in the estimation of probability, but that, as we shall see in our discussion of the relative-frequency theory, is chiefly a misconception of the meaning of these theorems. The calculus, as it stands, can only transform given probabilities and not create them.

It might be argued that answers to the questions we have posed were never intended in formulations of axiomatic probability and therefore it cannot be taken to task for failing to address these questions. In our pragmatic view this could only mean that these axiomatizations are seriously incomplete. The complexity, logical, and subjective theories offer guidelines for the assessment of probability. These theories take pains, for the most part, to provide enough of an indication of the nature of probability so that it can be estimated and, perhaps, intelligently used. It is in the expectation of learning more about probability theory and its meaning that we will study other theories of probability. First though we must examine the relation between probability and relative-frequency—a relation widely espoused as central to the formulation, justification, and measurement of probability.

References

1. A. Kolmogorov, *Foundations of the Theory of Probability*. Bronx, New York: Chelsea, 1956.
2. B. de Jouvenal, *The Art of Conjecture*, pp. 138–141. New York: Basic Books, 1967.
3. L. J. Savage, *Foundations of Statistics*, pp. 82–91. New York: Wiley, 1954.
4. E. B. Dynkin, *Markov Processes*, Vol. II, p. 201. Berlin and New York: Springer-Verlag, 1965.
5. R. von Mises and H. Geiringer, *The Mathematical Theory of Probability and Statistics*, pp. 58–74. New York: Academic Press, 1964.
6. M. Loeve, *Probability Theory*, p. 87. Princeton, New Jersey: Van Nostrand-Reinhold, 1960.
7. P. Suppes and J. Zinnes, Basic Measurement Theory, in *Handbook of Mathematical Psychology* (R. D. Luce, R. Bush, and E. Galanter, eds.), Vol. I, pp. 2–75. New York: Wiley, 1963.
8. D. Krantz, R. D. Luce, P. Suppes, and A. Tversky, *Foundations of Measurement*, Vol. I. New York: Academic Press, 1971.
9. J. Aczel, *Lectures on Functional Equations*, pp. 253–273. New York: Academic Press, 1966.
10. H. Cramér, *Mathematical Methods of Statistics*, p. 152. Princeton, New Jersey: Princeton Univ. Press, 1957.
11. K. Ito, Canonical Measurable Functions, *Proc. Int. Conf. Functional Anal. and Related Topics, Tokyo, April 1969*.
12. B. Gnedenko and A. Kolmogorov, *Limit Distributions for Sums of Independent Random Variables*. Reading, Massachusetts: Addison-Wesley, 1954.
13. J. Doob, *Stochastic Processes*, pp. 50–71. New York: Wiley, 1953.
14. A. Renyi, *Foundations of Probability*, p. 38. San Francisco, California: Holden-Day, 1970.
15. R. Cox, *The Algebra of Probable Inference*, pp. 4, 13–16. Baltimore, Maryland: Johns Hopkins Press, 1961.

IV

Relative-Frequency and Probability

IVA. Introduction

In Chapters II and III we discussed the rationale behind some of the formal structure of the concepts of random events, independence, and probability. We now turn to the substantive question of the interpretation of probability and its relation to the practical world. The view of probability that concerns us in this chapter is that it is a physical characteristic or description of the occurrence of events in the performance of an experiment. As a physical or empirical property of random events, probability should be objectively measurable from data and should be interpretable in terms of predictions about random events. More specifically we are interested in describing the pattern of occurrence of outcomes of what seem to be nondeterministically predictable experiments.

In Sections IVB and IVC we narrow the range of a physical interpretation of probability to the frequency of outcomes in an indefinitely repeated sequence of experiments and cast a new light on the basis for this interpretation. Sections IVD and IVE consider the formalizations of this concept due to Bernoulli/Borel and von Mises and the inadequacies of these formalizations. In Sections IVF, IVG, and IVH, we reach a negative assessment concerning the value of the formalized physical interpretation for either the measurement or utilization of probability. After a summation in Section IVI, we proceed in Section IVJ to outline three directions that have been followed in attempts to solve the probability measurement problem. Finally, in Section IVK we examine the possibility of a less ambitious undertaking—the measurement of comparative rather than quantitative probability.

IVB. Search for a Physical Interpretation of Probability Based on Finite Data

1. Reduction through Exchangeability

A physically useful interpretation of probability would presumably supply objective information about the occurrence of an event in the performance of an experiment. While other interpretations of probability attempt to make statements about the occurrence of an event in the performance of a single experiment, the relative-frequency approach essentially focuses on repeated experiments. We look for an invariant description of the outcomes of repeated experiments. Let us assume that we repeat an experiment $\mathscr{E}$ n times, with the individual performances $\mathscr{E}_1, ..., \mathscr{E}_n$ being unrelated or unlinked and having outcomes $x_1, ..., x_n$. For ease of presentation we will also assume that each outcome is either 0 or 1. What might constitute an invariant, based on the data $x_1, ..., x_n$, that is descriptive of the occurrences of 0 or 1 in a performance of $\mathscr{E}$?

The assumption that $\{\mathscr{E}_i\}$ are unlinked repetitions of the same experiment $\mathscr{E}$ implies that the desired description for the occurrence of events should be independent of the order of performance of the $\{\mathscr{E}_i\}$. The only functions of $x_1, ..., x_n$ invariant under permutation [i.e., $f(x_1, ..., x_n) = f(x_{i_1}, ..., x_{i_n})$ for any permutation $(i_1, ..., i_n)$ of $(1, ..., n)$] are functions of n and the number of occurrences of each outcome. In the case of $\{0, 1\}$-valued outcomes the description or probability p of the occurrence of $x = 1$ is such that $p = p(\sum^n x_i, n)$ or, equivalently, $p = p((1/n) \sum^n x_i, n)$.

To digress somewhat we note that in probabilistic terms the property of invariance under permutation of the order of performance of the experiments, used above, is weaker than that the experiments be unlinked repetitions of an experiment. Permutation invariance corresponds to assuming that the joint probability distribution of outcomes $P(x_1, ..., x_n)$ is a symmetric function; that is, for any permutation

$(i_1, ..., i_n)$ of $(1, ..., n)$, $P(x_1, ..., x_n) = P(x_{i_1}, ..., x_{i_n})$. Unlinked repetitions can be formalized through a joint distribution $P(x_1, ..., x_n)$ that factors as a product,

$$P(x_1, ..., x_n) = \prod_{i=1}^{n} P(x_i).$$

Clearly the latter implies the former but not conversely.

Random events or variables describable by symmetric joint distributions have been called exchangeable by de Finetti [1]. In the case of $\{0, 1\}$-valued random variables de Finetti characterized the class of symmetric n-variate distributions as mixtures of the following two types of n-variate distributions. Let $w = \sum_{i=1}^{n} x_i$:

   (i)   There exists a distribution function $F(p)$ with support $[0, 1]$,

$$P(x_1, ..., x_n) = \int_0^1 p^w (1 - p)^{n-w} \, dF(p);$$

   (ii) $(\exists w_0) \, P(x_1, ..., x_n) = \begin{cases} \binom{n}{w_0}^{-1} & \text{if } w = w_0 \\ 0 & \text{if } w \neq w_0. \end{cases}$

We may interpret type (i) exchangeable, binary-valued random variables as arising from Bernoulli trials with a randomly selected p. A result of later interest to us, proven in Loeve [2], is that $(1/n) \sum^n x_i$ converges (as n increases) with probability 1 to the random variable p having distribution function F. Type (ii) distributions are incompatible with the existence of a symmetric probability measure on the space of infinite sequences of random variables.

Returning to the search for a physically significant description of the occurrence of events in finitely many repetitions of an experiment, to be referred to as the probability of the event, we see that the often reasonable assumption that the order of performance of the experiments is irrelevant narrows our search to functions of the number n of repetitions and the relative-frequencies of occurrences for the possible outcomes. (These quantities also form a sufficient statistic with respect to estimation of parameters of exchangeable distributions.) If we fix n and agree that the probability of occurrence of an event depends on the data, then p will vary with the observed relative-frequency. However, experience suggests that there exist experiments, which we label as random, such that relative-frequency is not an invariant characteristic of the outcomes of repetitions of the experiment; if we take a fresh set of n repetitions of $\mathscr{E}$, then the new outcomes will usually have relative-frequencies that differ from those of the original data. We seem to be stymied in our attempt to locate a suitable invariant to describe the occurrence of outcomes of finitely many repeated experiments.

2. Maturity of Chances and Practical Certainty

Even if we could define probability in terms of the observed data, how would we use it? Knowing p could not tell us anything certain about yet to be performed repetitions of $\mathscr{E}$. Attempts to base probability on finite sequences of outcomes lead to the fallacy of the "maturity of chances." This fallacy suggests that if the probability of an event is p and in $k < n$ trials fewer than pk occurrences have been observed, then in the remaining $n - k$ trials there will be more than $p(n - k)$ occurrences; in this way the overall number of occurrences will balance near pn.[†]

It has been suggested that the fallacy of the maturity of chances can be avoided through the introduction of judgments of "practical certainty" to coordinate probability and a finite relative-frequency interpretation by determining when it is reasonable to expect the "maturity of chances" [3]. Practical certainty is taken to be an undefined primitive concept related to probability in that some events of "very high" probability, but no events of "low" probability, are practically certain to occur. If an event is judged practically certain to occur, then we act as if it will occur—the Cournot principle. Probability then scales the range from practical certainty of occurrence down to practical impossibility. Intermediate values of probability find meaning insofar as they enter, through the Kolmogorov calculus, into evaluations of the very high or very low probabilities. On this view, knowing $p = \frac{1}{2}$ we might conclude that there are practically certain to be between 30 and 70 heads in 100 unlinked tosses of a coin.

Of course, by introducing the notion of practical certainty we no longer have a strictly relative-frequency interpretation of probability or even, necessarily, an objective theory of probability. Yet we would agree that going beyond relative-frequency is inevitable in the search for an interpretation of probability. The problem though with the large role being presumed for practical certainty is that it too con-

---

[†] A class of random experiments for which maturity of chances and a finite relative-frequency characterization may be reasonable is the so-called urn models or sampling without replacement from a finite population. However, such models are but a small part of our interests in applied probability theory.

veniently lumps all of our difficulties with the interpretation of probability into one and then protects it from further examination on the grounds that it involves a primitive concept. Primitive concepts to be useful need to be clearer to our intuition than the grounds for judgments of practical certainty appear to be.

The sizable difficulties confronting attempts to measure probability from finite data and to forecast outcomes of a finitely repeated experiment given the probability, suggest that we look elsewhere for a definition of probability. Two possibilities are either to introduce infinitely repeated experiments or to define a probability that does not characterize or describe the actual outcomes of either a single or a repeated experiment.

IVC. Search for a Physical Interpretation of Probability Based on Infinite Data

1. Introduction

Finite relative-frequency has been argued to be an inadequate basis for probability. However, closer inspection of the behavior of relative-frequencies of empirical data sources or experiments suggests that they stabilize or converge as more and more experiments are performed. Relative-frequencies generally fluctuate in real examples but the fluctuations appear to diminish as the amount of data increases. The following quotations from respected probability texts illustrate this widely held view that relative-frequencies appear to converge and, therefore, that the limit of relative-frequency might provide the desired interpretation of probability.

> There exist today other examples of the verification of the abovementioned empirical fact [i.e., fluctuation of the relative-frequency takes place in the neighborhood of the probability of the event] which are of scientific and practical importance (Gnedenko [4, p. 54]).

> The fact that in a number of instances the relative frequency of random events in a large number of trials is almost constant compels us to presume the existence of certain laws, independent of the experimenter, that govern the course of these phenomena and that manifest themselves in this near constancy of the relative frequency (Gnedenko[†] [4, p. 55]).

[†] B. Gnedenko, *Theory of Probability*, 4th ed., pp. 54 and 55. Bronx, New York: Chelsea, 1968.

> *In spite of the irregular behavior of individual results, the average results of long sequences of random experiments show a striking regularity* (Cramér[†] [5, p. 141]).

This observed regularity, that the frequency of appearance of any random event oscillates about some fixed number when the number of experiments is large, is the basis of the notion of probability (Fisz[‡] [6, p. 5]).

> *A random (or chance) phenomenon* is an empirical phenomenon characterized by the property that its observation under a given set of circumstances does not always lead to the same observed outcome (so that there is no deterministic regularity) but rather to different outcomes in such a way that there is *statistical regularity* (Parzen[§] [7, p. 2]).

We may sum up the import of the above quotations in the following points:

(1)   In many important cases relative frequencies appear to converge or stabilize when one sufficiently repeats a random experiment.

(2)   Apparent convergence of relative-frequency is an empirical fact and a striking instance of order in chaos.

(3)   The apparent convergence suggests the hypothesis of convergence and thereby that we may extrapolate from observed relative-frequency to the relative-frequency of outcomes in as yet unperformed trials of an experiment.

(4)   Probability can be interpreted through the limit of relative-frequency and assessed from relative-frequency data.

We undertake to explain (1) in such a manner that (2) and (3) are seriously challenged. Once (2) and (3) are challenged, (4) is still possible but not particularly interesting. Our explanation of (1) is achieved by proving that a reasonable definition of a random sequence implies that all such sequences must have apparently convergent relative-frequencies. This end is accomplished by adopting the computational-complexity approach to random sequences as pioneered, in a sense, by von Mises and Church and as brought to its present state by recent efforts of Chaitin, Kolmogorov, Solomonoff, and Martin-Löf (see Chapter V). In Subsection IVC2 we formally define what we mean by the "apparent convergence" of relative-frequency and establish a scale by which to measure the "degree of randomness" of a sequence. For simplicity and clarity the remainder of this discussion will focus exclusively on

---

[†] H. Cramér, *Mathematical Methods of Statistics*, p. 141. Princeton, New Jersey: Princeton Univ. Press, 1957.

[‡] M. Fisz, *Probability Theory and Mathematical Statistics*, 3rd ed., p. 5. New York: Wiley, ©1963, by permission of John Wiley & Sons, Inc.

[§] E. Parzen, Modern Probability Theory and Its Applications, p. 2. New York: Wiley, ©1960, by permission of John Wiley & Sons, Inc.

binary sequences $S_n = (x_1, ..., x_n)$ in which x_i is either 0 or 1. Subsection IVC3 develops a basic theorem relating the two concepts formalized in Subsection IVC2. In Subsection IVC4 we apply the theorem of Subsection IVC3 to explain points (1), (2), and (3).

2. Definition of Apparent Convergence and Random Binary Sequence

The phrase "apparent convergence of relative-frequency" is meant to apply to the seeming increasing stability of relative-frequency as the number of trials increases but remains finite. A straightforward definition of apparent convergence, involving the introduction of two parameters (m, ϵ) to determine the point at which convergence is expected to be exhibited, m, and the degree of fluctuation, ϵ, is provided by the following.

Definition. The sequence $r_1, ..., r_n$ apparently converges (m, ϵ) if

$$\max_{m \leqslant j \leqslant n} |r_j - r_n| < \epsilon.$$

For a sequence $(x_1, ..., x_n)$ of $\{0, 1\}$-valued outcomes the relative-frequency based on the first j terms, r_j, is given by

$$r_j = \frac{1}{j} \sum_{k=1}^{j} x_k.$$

The choice of parameters (m, ϵ) is left open. Clearly small m and small ϵ represent the more stringent conditions for apparent convergence.

The definition of a random sequence $S_n = (x_1, ..., x_n)$ is a more involved question and is discussed more fully in Chapter V. From one point of view a random sequence is maximally irregular or complex. Looking ahead to the more complete discussion contained in Sections VC and VD, we assert that a suitable measure of the complexity $K_\Psi(S \mid I)$ of a sequence S, given information I, is the length $l(P)$ of the shortest program P, which when combined with the input I, generates S on some appropriate computing machine Ψ,

$$K_\Psi(S \mid I) = \min\{k : k = l(P), \Psi(P, I) = S\}.$$

In Sections VE and VG we substantiate that an approach based on this measure of complexity yields classifications of sequences as random or nonrandom which are consistent with classifications by statistical tests for randomness.

In assessing whether a sequence S is random or not and, therefore, whether we wish to apply a relative-frequency-based theory of probability (exchangeable model) to the description of S, we assume that we have as side information the length of S, $l(S)$, and the relative-frequency of 1's in S or, equivalently, the weight w of S,

$$w(S) = \sum_{i=1}^{l(S)} x_i \, .$$

Hence our test for randomness will depend on $K_\Psi(S \mid l(S), w(S))$ for some suitable computer (AUTM)Ψ.

To set a threshold for the degree of complexity to be judged as random we restate results from Section VC.

Theorem 1. Ψ an AUTM

$$(\exists c)(\forall S) \, K_\Psi(S \mid l, w) < \log_2 \binom{l}{w} + c.$$

There are fewer than $2^{-t}\binom{l}{w}$ sequences S for which

$$K_\Psi(S \mid l, w) < \log_2 \binom{l}{w} - t.$$

Hence, most sequences have complexity at least $\log_2\binom{l}{w} - t$ and none have complexity greater than $\log_2\binom{l}{w} + c$. This observation suggests, and the work in Section VG strongly confirms, the reasonableness of the following.

Definition.   The binary sequence S of length l and weight w is random (Ψ, t) where Ψ is an AUTM, if

$$K_\Psi(S \mid l, w) > \log_2 \binom{l}{w} - t.$$

Clearly, the smaller t is the more stringent the definition of randomness becomes. The choice of AUTM Ψ is left open.

3. Relations between Apparent Convergence of Relative-Frequency and Randomness

The key result relating apparent convergence (m, ϵ) and randomness (Ψ, t) is given by

Theorem 2.

$$(\exists c)(\forall m, \epsilon > 0)(\forall S) \, K_{\Psi}(S \mid l, w) > \log_2 \left(\begin{array}{c} l \\ w \end{array} \right) - \log_2 (m\epsilon^4) + c$$

$$\Rightarrow \max_{m \leqslant j \leqslant l} \left| \frac{1}{j} \sum^{j} x_i - \frac{w}{l} \right| < \epsilon.$$

Proof.   See Fine [8, Appendix].   ∎

In words, a sequence of high complexity, (Ψ, t) random where $t = \log_2(m\epsilon^4) - c$, is one for which the relative-frequency appears to converge (m, ϵ).

Reflection upon Theorem 2 leads us to several conclusions regarding the import of the empirical observation that relative-frequency appears to stabilize in the long run. The first conclusion we reach is that the common assumptions that convergence of relative-frequency conflicts with irregularity, and that in irregular sequences of interest relative-frequency only fortuitously appears to stabilize, are the opposite of the truth. Apparent convergence of the relative-frequency occurs because of, and not in spite of, the high irregularity (randomness) of the data sequence!

The second conclusion is that contrary to our quotations from Gnedenko, stability of relative-frequency is not so much an empirical fact reflective of a law of nature as the outcome of our approach to data. An examination of our data-processing practices in science and engineering suggests that if we are aware that a data sequence S with parameters (l, w) can be described on some computer by a program P with

$$l(P) \ll \log \left(\begin{array}{c} l \\ w \end{array} \right),$$

then we will not attempt to characterize S through the relative-frequency w/l of outcomes in S. Rather we would make use of the structure in S implied by the existence of a short descriptive program and not treat the individual outcomes as if they were independent and identically distributed. The point is that those data sequences S to which we commonly apply the relative-frequency argument have failed tests for structure; they are of high complexity. It unexpectedly follows from Theorem 2 that those sequences that we select for application of the relative-frequency approach have been simultaneously preselected to possess apparently convergent relative-frequencies. It is the scientist's selectivity in using the relative-frequency argument that accounts for his success, rather than some fortuitous law of nature.

The third conclusion, and one of significant practical importance, is that since the apparently convergent relative-frequencies in an empirical random sequence do not arise from, and are not reflective of, an underlying empirical law, we have no grounds for believing that the apparent limit is indicative for future data. Convergence was forced by preselection, on grounds once thought to conflict with convergence, of the data sequence and is not a reflection of some underlying process of convergence governing all trials. The appearance of stability is not, as it has been thought, evidence for the hypothesis that nature ensures stability. Our intuitive expectations that apparent trends must persist has even less justification than would appear to a skeptic familiar with Hume's firm attack on induction; the trends are of our own making.

To balance these observations, we should note though that we are often negligent in our search for structure and may assume S to be of high complexity when it is not (e.g., pseudorandom sequences). Our argument for the stability of relative-frequency only applies when the complexity of S is at least $\binom{l}{w} - O(\log l)$. Stability of the relative-frequency in less complex sequences is contingent rather than necessary.

4. Conclusion

Granting that the appearance of stability of the relative-frequency is deceptive, can we not yet assert that it is true? Such an assertion would be metaphysics and subject neither to experimental proof nor to disproof. Observation of any finite number of experiments would tell us nothing about limits of relative-frequencies. Objectivity and rationality should make us exceedingly reluctant to adopt and rely on such a metaphysical basis. Our experience is finite, not infinite, and our concerns are with finite repetitions and not infinite repetitions of experiments. We agree with a recent remark of Kolmogorov[9],

> The frequency concept based on the notion of limiting frequency as the number of trials increased to infinity, does not contribute anything to substantiate the applicability of the results of probability theory to real practical problems where we have always to deal with a finite number of trials.

Yet for the sake of discussion, and in an attempt to better understand claims tendered in favor of a limit of relative-frequency interpretation of probability, let us provisionally adopt this position. We now inquire

---

[†] A. Kolmogorov, On Tables of Random Numbers, *Sankhyā Ser. A*, p. 369, 1963, by permission of Statistical Publishing Society.

into the relation between the Kolmogorov and von Mises theories of probability and the limit of a relative-frequency interpretation, how this interpretation assists us in measuring probability, and how it enables us to go from given probabilities to statements about outcomes of repeated experiments.

IVD. Bernoulli/Borel Formalization of the Relation between Probability and Relative-Frequency: Strong Laws of Large Numbers

The Bernoulli/Borel ensemble formalization of the relation between relative-frequency and probability, if we conveniently restrict ourselves to binary-valued experiments, is that

$$P\left(\lim_{n\to\infty}\frac{1}{n}\sum_i^n x_i = p\right) = 1.$$

In words, it is asserted that with overwhelming probability, but not in all cases, there is a limit for the sequence of finite relative-frequencies and this limit is identical to the probability of occurrence of the event. When the indefinitely repeated, unlinked experiments are formalized as independent and identically distributed outcomes, this relation between probability and relative-frequency becomes a theorem known as a strong law of large numbers (SLLN).

The role of the formalization of repeated, unrelated, or unlinked experiments as having stochastically independent and identically distributed (IID) outcomes is often understated by commentators on the strong laws as presumed bridges between theory and practice. As we noted in Section IVB, if we model such experiments as giving rise to exchangeable outcomes, a weaker and therefore more readily defensible interpretation, then we can still conclude that $(1/n)\sum^n x_i$ converges with probability 1. However, the limit is now a random variable and not necessarily degenerate at the value $P(x_i = 1)$. It is, of course, true that there are other conditions than IID under which $\{x_i\}$ has a convergent relative-frequency—ergodicity [2, Chapter IX] being the primary statement. However, these other formal conditions guaranteeing convergent relative-frequency are much harder to identify from an informal or intuitive understanding of the interrelations in a sequence of experiments.

It is only insofar as our informal or intuitive perception of the hypotheses of a SLLN is sufficiently developed to enable us to judge when they apply to a particular sequence of experiments that we can think of crossing over from the experiment, via the SLLN, to an expectation that the relative-frequency of outcomes will be stable.

What though have we achieved when in a given case, through some obscure legerdemain, we correctly assume a formalization of the experimental setup that enables us to prove a SLLN? We are still unable to assert that the particular sample sequence being generated by our actual experiment is in fact one having a relative-frequency converging to a probability p. The SLLN only says that the set C_p of such sequences has a probability of 1. The complementary set $\bar{C}_p$ of sequences $\{x_i\}$ for which $(1/n) \sum^n x_i$ does not converge to $P(x_i = 1) = p$ is not a null set. In fact there is a one-to-one correspondence between the sequences in C_p and in $\bar{C}_p$. The set of sample sequences for which the relative-frequency does not converge to the probability is as large a set, in the sense of cardinality, as the set for which it does converge. Furthermore, the set D of sample sequences for which the relative-frequency diverges,

$$D = \overline{\bigcup_{0 \leqslant p \leqslant 1} C_p},$$

also has the same cardinality as the set for which the relative-frequency converges to the probability. The verification of the equality of cardinality of the sets of sequences C_p and $\bar{C}_p$ follows from the fact that C_p has Bernoulli p-measure 1 and $\bar{C}_p$ has Bernoulli p'-measure 1 for $p' \neq p$; both sets, for discrete x, have the cardinality of the real line. The set of sequences with divergent relative-frequencies is also uncountable and therefore has the cardinality of the real line as well.

Thus we cannot blithely ignore the set D or $\bar{C}_p$ just because the SLLN shows that for a particular measure (Bernoulli p) they have probability 0. Probability 0 does not necessarily mean a small or negligible set of possibilities. All infinite sample sequences in any Bernoulli p-measure for $0 < p < 1$ have the same probability of 0. Hence, we cannot even argue that some infinite sequences, and we can only talk about limits of relative-frequency for infinite sequences, are more probable than others and thereby skirt the import of the equality of cardinality. There are as "many" ways to get divergent relative-frequencies as there are to get convergent relative-frequencies and all the ways are equally probable.

In practice we require an additional assurance that the sequence we are observing is one in a set C_p rather than in D or $C_{p'}$. One might be tempted to argue at least that if we repeat the overall experiment, then most of the cases will produce sequences in C_p—that is, we apply the relative-frequency interpretation to the probability 1 guarantee provided by the SLLN. It is easily seen though that this argument is open to precisely the same objections even though it removes the source of difficulty to another level (a superexperiment having as its outcomes an infinite sequence); we no longer require a guarantee that each sequence has a convergent relative-frequency but only that in practice most of them do. We cannot circumvent the need for an assurance at some level that a sequence we are generating is actually in a convergence set. This difficulty is reminiscent of the often lightly posed but nevertheless serious query, "How do you know that we aren't living in a world of zero probability?"

IVE. Von Mises' Formalization of the Relation between Probability and Relative-Frequency: The Collective

The Bernoulli/Borel formulation relied upon an ensemble approach. It was only asserted that the relative-frequency of the outcomes of an event in n unlinked repetitions of an experiment converged to the probability of the event with probability 1; in the ensemble of all sequences there was a subset of convergent sequences of measure 1. It was then possible to prove this statement (SLLN) through a formalization of repeated unlinked experiments as being IID outcomes. As we pointed out in the preceding section, a "with probability 1" guarantee is surprisingly weak. Von Mises straightforwardly confronts this problem by postulating that the actual sample sequence, which he called a *collective*, has a convergent relative-frequency. Von Mises also goes further than Bernoulli/Borel by requiring that the sample sequence be "random" before he is willing to supply a relative-frequency-based probabilistic model for it. The judgment of "randomness" is made in accordance with an interesting, operational criterion that is free from probabilistic considerations. It is on these two grounds that we prefer the von Mises approach, sketched below, to the Bernoulli/Borel approach.

Before stating von Mises' formalization we need to define the operation of subsequence generation by place-selection functions. It is this operation that formalizes the idea of unlinked repeated experiments in terms of the actual sample sequence rather than in terms of the less accessible probability distributions. Let Ω contain M possible outcomes

ω_1 , ..., ω_M . We uniquely and invertibly map the sequence of the first n outcomes x_1 , ..., x_n into a positive integer s_n , say through

$$s_n = \sum_{k=1}^{n} i_k (M + 1)^{k-1},$$

where i_k is such that $x_k = \omega_{i_k}$. A place-selection function ϕ, following Church [10], is an effectively computable (i.e., it can be computed by a Turing machine) binary-valued (say $\{0, 1\}$) function of a positive integer argument. We employ ϕ to generate a subsequence S' of the infinite sequence of outcomes $S = x_1, x_2, \ldots$ as follows: x_{n+1} is in S' if $\phi(s_n) = 1$ and x_{n+1} is not in S' if $\phi = 0$.

We may now follow von Mises and Geiringer [11] and define the type of sequence, called a collective, for which it may be reasonable to hold a limit of a relative-frequency interpretation of probability.

Definition. Collective $K(\Phi, \Omega)$ is an infinite sequence S of elements of Ω and a countable family Φ of place-selection functions that is closed under composition (i.e., if ϕ' applied to S yields S' and ϕ'' applied to S' yields S^*, then there is a ϕ^* which when applied to S yields S^*) such that

(1) $(\forall \omega_i \in \Omega) \lim_{n \to \infty} N_{\omega_i}(n)/n = p_i$ [$N_{\omega_i}(n)$ is the number of occurrences of ω_i in the first n terms of S].

(2) $\sum_{i=1}^{\infty} p_i = 1$.

(3)   If we generate an infinite subsequence S' from S by means of any $\phi \in \Phi$, then S' satisfies (1) and (2) for the same $\{p_i\}$.

The choice of Φ may depend on the problem under consideration or it may be the countable set of all effectively calculable Φ. It should be observed that the collective contains only a single infinite sequence, in contrast to the usual ensemble of sample functions setup. Condition (1) asserts the existence of the limiting relative-frequency for each possible experimental outcome $\omega_i \in \Omega$. Condition (2), corresponding to the unit probability normalization, is unnecessary if Ω is a finite set but is necessary when Ω is infinite. In the von Mises analysis the unit probability normalization asserts something about the outcome of repeated experiments beyond that contained in the existence of relative-frequencies. Condition (3) incorporates a notion of a random sequence and has no counterpart in the formal Bernoulli/Borel/Kolmogorov system. It is an attempt to assert the nonexistence of an unfair gambling scheme; any scheme for selecting favorable trials on which to wager can be described by a place-selection function ϕ that generates a subsequence having the same frequency of "wins" as in the full sequence. A more detailed and broader exposition of the von Mises theory is available in [11]. In particular, we omit discussion of the basic ways of generating new collectives from a given collective which lead to definitions of conditional probability and independence and to proofs that the collective sequence has many of the properties we expect of a random sequence (e.g., all k-tuples of elements in Ω give rise to a new collective).

Interestingly, Ville [12] has pointed out that the von Mises–Church notion of a collective relative to even the whole set of effectively computable place-selection functions includes, say, binary sequences in which all initial segments contain at least as many 1's as 0's. In gambling terms, there are unfair collectives in which the lead never changes and

$$\sup_{n} \sum_{i=1}^{n} (x_i - \tfrac{1}{2}) = \infty.$$

This conflicts with our expectation that in a random sequence the lead should change infinitely often. Not only should no player be able to improve his winnings by clever wagering, no player should be able to stay ahead indefinitely.

The concept of a collective can be strengthened, so as to eliminate the preceding difficulty and some others like it, by a natural extension of the von Mises principle of the impossibility of a gambling scheme that was first suggested by Ville and which led to his development of the theory of martingales. A gambling system is a function,

$$\psi : \{0, 1\} \times Z^+ \to [0, 1],$$

that associates to each possible outcome $\omega \in \{0, 1\}$, each history $s_n \in Z^+$ (positive integers), and fortune F_n, a stake $\psi(\omega, s_n) F_n$ on that outcome.

The cumulative fortune F_n following trial n is given by

$$F_n = 2\psi(x_n, s_{n-1})F_{n-1}.$$

The requirement that the infinite sequence S of outcomes be random relative to the system ψ is that

$$\sup_{n} F_n < \infty.$$

Hence we might adopt the following.

Definition. The infinite sequence S is random if, when we denote the set of all effectively computable gambling systems by Ψ,

$$\sup_{\psi \in \Psi} \sup_n F_n < \infty.$$

In other words, no effectively computable gambling system can yield arbitrarily large fortunes when the sequence of outcomes is random.

The notion of an infinite random sequence is examined from the viewpoint of computational-complexity in Chapter V. Kolmogorov [13] has noted the existence of collectives for which the complexity of initial segments of length n is $O(\log n)$ and not $O(n)$ as required by the newer definitions given in Section VF. Notwithstanding its defects, von Mises' suggestion that it is only for collectives that probability and its relative-frequency interpretation are reasonable is a substantial addition to the Kolmogorov axiomatic theory. A guideline is provided for the recognition of the domain of probability that renders it distinguishable from measure. Of course, in practice, it would be impossible to recognize whether a finite sample sequence is indeed the initial segment of a collective.

IVF. Role of Relative-Frequency in the Measurement of Probability

If, despite the shortcomings of the Bernoulli/Borel/Kolmogorov and von Mises formulations, we accept the fact (i.e., metaphysical assumption) that we are observing the first few terms $x_1, ..., x_n$ in an infinite sequence S of independent and identically distributed random variables, then how can we use the data $x_1, ..., x_n$ to measure or assess $P(x_1 = 1) = p$? Looked at from the viewpoint of the Kolmogorov formulation, a minimal requirement that a function $\hat{p}(x_1, ..., x_n)$ be a measurement or estimate of p is that

$$(\exists \epsilon < 1, \delta > 0)\, P(|\, p - \hat{p}(x_1, ..., x_n)| < \epsilon \mid x_1, ..., x_n) \geqslant \delta.$$

A trivial example of an estimate is the constant function $\hat{p} = \frac{1}{2}$ for $\epsilon \geqslant \frac{1}{2}$. If we assume that p is unknown, but not random, then $P(|\, p - \hat{p}\,| < \epsilon \mid x_1, ..., x_n)$ is either 0 or 1. If we omit the trivial cases based on the certainty that $0 \leqslant p \leqslant 1$, then there does not exist any estimate $\hat{p}$ of p. If we assume that p is random with a known prior distribution (S is exchangeable), then there exist estimators $\hat{p}$ that are functions of the relative-frequency, satisfying the above-mentioned minimal requirement of measurement for nontrivial ϵ and δ. However, since the prior for p is either not data based or else depends on other than frequency data, the resulting estimator $\hat{p}$ is either not fully empirical or else depends on more than just the explicit frequency data. In either case, relative-frequencies alone do not yield probability measurements.

It is of course true that even for nonrandom p and unlinked experiments we can find $\hat{p}((1/n) \sum^n x_i , n)$ such that

$$(\forall \epsilon > 0)(\exists \delta > 0)\, P(|\, p - \hat{p}\, | < \epsilon) > \delta.$$

For example, if we take $\hat{p} = (1/n) \sum^n x_i$, then application of the Tchebychev inequality easily yields

$$P\left(\left|\, p - \frac{1}{n}\sum_{}^{n} x_i \right| < \epsilon\right) \geqslant 1 - \frac{p(1-p)}{n\epsilon^2} \geqslant 1 - \frac{1}{4n\epsilon^2}\,.$$

However, in any given case we are not interested in the measurement accuracy "averaged" over an ensemble of possible observations but in the accuracy of the measurement based on the actual observations.

If we now examine matters as they appear using the von Mises collective, we are not significantly better off. The probability p is now identified exactly with

$$\lim_{n\to\infty} \frac{1}{n}\sum_{}^{n} x_i \,.$$

However, we have absolutely no information as to the rate of convergence. For no n, however large, and $\epsilon < \frac{1}{2}$ (to avoid trivialities) can we assert that $|(1/n) \sum^n x_i - p\,| < \epsilon$. There is no uniformity of convergence for the family of collectives. Once again we are unable to provide an estimate $\hat{p}$ of p based on finite data and satisfying very weak requirements as to what a measurement or estimate is.

Note the contrast between the problem of measuring probability and that of measuring the usual physical quantities such as resistance. In the case of resistors we can if we wish guarantee to estimate the resistance of a resistor R to within the larger of 100 ohms or 20% of the measured resistance R. We can supply a finite region

$$|\, \hat{R}(x_1 , ..., x_n) - R\, | < \epsilon$$

in which the true quantity is certain[†] to lie. This is not the case with probability interpreted as relative-frequency.

[†] By "certainty" we mean the certainty with which we can determine the experimental outcome ω; for example, the certainty with which we know "head" was the outcome of an observed coin toss. We distinguish between this level of certainty and that with which we make predictions about as yet unobserved experiments.

In brief, even with the relative-frequency interpretation, we cannot arrive at probability conclusions without probability antecedents. In the absence of an empirical, frequency-based prior, there can be no empirical frequency-based posterior.

IVG. Prediction of Outcomes from Probability Interpreted as Relative-Frequency

Putting aside the apparent difficulties in obtaining probability from data, of what use would such probabilities be if we could estimate them? What prediction could we make regarding the outcome $x_1, ..., x_n$ of unlinked, repeated experiments if we knew $p = P(x_i = 1)$?

In the Kolmogorov formulation we could of course give the probability of any outcomes,

$$P(x_1, ..., x_n) = p^{\Sigma x_i}(1 - p)^{n - \Sigma x_i}.$$

This probability presumably tells us that with probability 1 if we indefinitely repeat the compound experiment, $\mathscr{E} = (\mathscr{E}_1, ..., \mathscr{E}_n)$, then the frequency with which the outcome $\{x_1, ..., x_n\}$ occurs approaches the known $P(x_1, ..., x_n)$. However, as we pointed out in discussing the SLLN, this statement is of no practical value. We would need a separate assurance that our sample sequence of compound experiments is in a convergence set. Furthermore, even if we had this assurance, it does not tell us what will happen to $x_1, ..., x_n$, the experiment we actually perform, but rather what will happen in a hypothetical indefinite repetition of this experiment.

Even with the assurance that

$$\lim_{n \to \infty} \frac{1}{n} \sum_{}^{n} x_i = p,$$

we are unable to say anything about $x_1, ..., x_n$. The events

$$\left\{ \lim_{n \to \infty} \frac{1}{n} \sum_{}^{n} x_i = p \right\} \quad \text{and} \quad \{x_1, ..., x_n\}$$

are independent—knowing that the first occurs tells us nothing about the second.

We can restate the preceding objection from the von Mises viewpoint. Knowing p tells us that $\lim(1/n) \sum^n x_i = p$. However, the absence of uniform convergence over the family of collectives and a lack of information on the rate of convergence of the actual sample sequence leaves us unable to come to a conclusion about $\{x_1, ..., x_n\}$ from p. All outcomes are compatible with p. Thus a relative-frequency interpretation for p in either the Bernoulli/Borel/Kolmogorov or von Mises formulations does not enable us to predict properties of actual experimental outcomes.

IVH. The Argument of the "Long Run"

While we have argued that knowing $\{(1/n) \sum^n x_i\}$ converges neither enables us to measure (estimate) the limit p nor does knowing p enable us to infer properties of $x_1, ..., x_n$, nevertheless the (metaphysical) hypothesis of convergence assures us that in the "long run" ("large" n) $\{(1/n) \sum^n x_i\}$ will eventually approximate to p. This is the basis of the usual "long run" justification of a limit interpretation of probability. The obvious difficulty with this argument is that there is no presumed uniformity of convergence over the set of sample sequences, and hence we do not *a priori* know how large "large" n must be for $|(1/n) \sum^n x_i - p| < \epsilon$. This situation would be somewhat more tolerable if it could be asserted that once we actually are within ϵ of the limit we can recognize it, although we could not determine the required n in advance of observation. For example, while I do not know how long it will be before, if ever, I visit the moon, I can at least determine it when this event occurs. The long run guarantee of convergence is thus ineffective since we are unable, either before or after observation, to determine how long a "long" run must be. (See Subsection IVK3.)

Finally, we join with Lord Keynes in his reputed rejoinder concerning the practical value of the long run justification: "In the long run we shall all be dead."

IVI. Preliminary Conclusions and New Directions

Our discussion to this point has proceeded from the search for a physical/objective interpretation of probability, relating it to a description of the occurrence of events in the performance of "random" experiments, to examination and rejection of finite relative-frequency and the limit of a relative-frequency as formalized by Kolmogorov and von Mises, and finally to a critical inspection of the interplay of probability and relative-frequency when we estimate probability from data and predict data from probability. In our view, the preceding sections have cast doubt upon the value of relative-frequency for the description of the occurrence of events.

Perhaps we are asking for too detailed a description of random experiments. How do we know that the specificity of quantitative probability corresponds to any interesting real random experiment? The analysis of the Kolmogorov axioms given in Chapter III did not really uncover a justification for quantitative, as opposed to comparative, probability. This point may gain in impact from the following pseudo-psychological reflections. How do we know, or why do we believe, that we cannot accurately predict the outcome of a vigorous toss of a balanced coin? I submit that this belief is primarily a product of personal frustration with a lack of success in actual attempts and secondarily an acceptance of the judgment of others that it cannot be done successfully. (Researchers into extrasensory phenomena and gamblers, though, may not share these beliefs.) Giving up on deterministic prediction of the outcomes of coin tosses, we then look for weaker but still informative predictions. Can we, for example, put narrow bounds on the exact number of heads in n tosses? Apparently our personal experiences indicate that we cannot correctly assert that "there will be between 40 and 60 heads in 100 tosses." So we proceed yet further along the path of looser predictions and talk about what will happen in a hypothetical long run. At this point we find ourselves making useless but safe predictions. Perhaps what we are attempting is not possible. We persist because the indefiniteness of our long run predictions makes them nearly irrefutable and reduces the incidence of frustration.

In subsequent sections we consider two paths out of the impasse discussed above. One path is to define axiomatically desirable characteristics of a probability estimate based on occurrences of events. In this way we solve the measurement problem but are left with a question as to the value of the estimated probability. In effect we define the characteristics a unicorn must have until we have a complete description of the beast—if it exists. The other path of induction by enumeration leads us to ask about the measurement of comparative rather than quantitative probability. We preserve the empirical core of relative-frequencies but do not attempt to derive, more or less arbitrarily, quantitative probabilities.

IVJ. Axiomatic Approaches to the Measurement of Probability

1. Introduction

An axiomatic approach specifies desirable properties of an estimator of a probability distribution. It is possible in this fashion to define what the estimator should be, even though the quantity being estimated may not exist. A simple illustration of the fallacy of such reasoning is the following: If we assume (incorrectly) that $\sin x = ax + b$, then evaluation at $x = 0, \pi$ proves that $a = b = 0$. If we disregard this reservation, then we can distinguish three classes of axiomatic approaches. The first class defines an estimator as a function of the data without reference to an explicit probabilistic model for either the data or the unknown probability. The second class (statistical decision theory) utilizes a probabilistic model for the data source as well as axioms to define what is meant by an estimator of probability. Finally, the third class (Bayesian statistics) assumes the unknown probability to have a prior distribution and that there is some probabilistic model for the data; the properties of the prior distribution as well as those of an estimator are then axiomatized. In the following subsections we provide an illustration of each approach, without attempting a survey [14].

2. An Approach without Explicit Probability Models

The following axioms for the estimator $\{\hat{p}_i\}$ of the probabilities $\{P(\omega_i)\}$, modeled on work of Johnson and Salmon [15], seem reasonable when the data are judged to consist of outcomes from an exchangeable sequence of experiments, each experiment having k possible outcomes $\omega_1, ..., \omega_k$.

A1. $\{\hat{p}_i\}$ are functions of only the number of occurrences $n_1, ..., n_k$ of each possible outcome:

$$\hat{p}_i(x_1, ..., x_n) = h(i, n_1, ..., n_k).$$

A2. The estimator $\hat{p}_i$ of $P(\omega_i)$ does not depend on the specific values of the numbers of other outcomes n_j $(j \neq i)$:

$$\hat{p}_i(x_1, ..., x_n) = f\left(i, n_i, \sum_{j=1}^{k} n_j, k\right).$$

A3. *A priori* there is no distinguished outcome:

$$f\left(i, n_i, \sum_{j}^{k} n_j, k\right) = g\left(n_i, \sum_{j}^{k} n_j, k\right).$$

A4. The estimator $\{\hat{p}_i\}$ is itself a probability distribution:

$$g\left(n_i, \sum_{j}^{k} n_j, k\right) \geqslant 0, \qquad \sum_{i=1}^{k} g\left(n_i, \sum_{j}^{k} n_i, k\right) = 1.$$

A5. The estimator $\hat{p}_i$ of $P(\omega_i)$ does not depend on the number of possible outcomes k:

$$\hat{p}_i(x_1, \ldots, x_n) = \hat{p}\left(n_i, \sum_{j}^{k} n_j\right).$$

Axiom A1 is a consequence of exchangeability. For any permutation $(i_1, \ldots, i_n)$ of $(1, \ldots, n)$ we should have $\hat{p}_i(x_{i_1}, \ldots, x_{i_n}) = \hat{p}_i(x_1, \ldots, x_n)$. The assertion of A1 immediately follows.

Axiom A2 is based on the belief that the occurrences of a given outcome ω_1 are uninfluenced by exactly how many times each of the other outcomes occur. This axiom is supplemented by A5, discussed below.

Axiom A3 incorporates the assumption that there is a balance of prior information regarding the $P(\omega_i)$; what is known about $P(\omega_i)$, prior to the data, applies as well to $P(\omega_j)$. Hence we should form estimates for $P(\omega_i)$ in the same manner as for $P(\omega_j)$.

Axiom A4 seems reasonable from a conventional statistical viewpoint. Knowing that a parameter $\theta \in \Theta$, we would usually estimate it, by some $\hat{\theta} \in \Theta$. The estimate of a probability distribution should itself be a probability distribution.

Axiom A5 derives from the belief that the occurrences of a given outcome ω_1, measured by $P(\omega_1)$, are uninfluenced by how many other kinds of outcomes we assume to be possible. We should be able to hypothecate new possible outcomes or combine several outcomes into one without then having to change the estimate $\hat{p}_i$ of $P(\omega_i)$.

It has been proposed that portions of A1–A5 can be defended by appeal to the principle of linguistic invariance enunciated by Salmon[†] as follows:

> If (1) e and e' are any two evidence statements in the same or different languages,
> (2) h and h' are two hypotheses in the same languages as e and e', respectively,

[†] W. C. Salmon, *The Foundations of Scientific Inference*, p. 102, by permission of the University of Pittsburgh Press. ©1966 by the University of Pittsburgh Press.

and (3) e is equivalent to e' and h is equivalent to h' by virtue of the semantical and syntactical rules of the languages in which they occur, then the inductive relation between e and h must be the same as the inductive relation between e' and h' [15, p. 102].

However, the interpretation and utility of this principle have been challenged by philosophers concerned with induction.

Turning to the implications of our axioms, we note the following.

Lemma. The functions $g(j, n, k)$ satisfying A1–A4 for $n > k > 2$ and $0 \leqslant j \leqslant n$ are such that for any $1/n \leqslant g(1, n, k) \leqslant (n-1)/n(k-1)$,

$$g(j, n, k) = g(1, n, k) - (j-1)\frac{kg(1, n, k) - 1}{n - k}.$$

Proof. Note that if for $k > 2$, we hypothetically combine ω_1 and ω_2 into a single outcome, then from A1–A3,

$$g(j_1, n, k) + g(j_2, n, k) = g(j_1 + j_2, n, k - 1).$$

Letting $j_2 = 0$ enables us to express $g(j_1, n, k - 1)$ as $g(j_1, n, k) + g(0, n, k)$ and yields the functional identity

$$g(j_1, n, k) + g(j_2, n, k) = g(j_1 + j_2, n, k) + g(0, n, k).$$

Finite induction confirms that the unique solution to this identity is

$$g(j, n, k) = [g(1, n, k) - g(0, n, k)]\, j + g(0, n, k).$$

If we now apply the unit normalization of A4, we find

$$g(j, n, k) = g(1, n, k) - (j-1)\frac{kg(1, n, k) - 1}{n - k}.$$

The nonnegativity requirement of A4 leads to the condition

$$\frac{1}{n} \leqslant g(1, n, k) \leqslant \frac{n - 1}{n(k - 1)},$$

and completes the proof of the lemma. ∎

If we now add A5 to A1–A4, we easily find the unique solution

$$\hat{p}_i(x_1, \ldots, x_n) = \frac{n_i}{n}.$$

Thus A1–A5 provide an axiomatic basis for finite relative-frequency as an estimator of the probability of occurrence of an event.

3. The Approach of Statistical Decision Theory

A great deal has been written on a variety of statistical principles to obtain probability estimates from data [5, 16]. The brief example we provide is meant to be illustrative and reasonable although not the "best" approach; there is no specific best answer from the viewpoint of statistical decision theory. We model the $\{0, 1\}$-valued outcomes $x_1, ..., x_n$ by the nonparametric hypothesis that they are type (i) exchangeable. From Subsection IVB1 we know that this implies there exists some distribution function $F(p)$ such that

$$P(x_1, ..., x_n) = \int_0^1 p^{\sum^n x_i}(1 - p)^{n - \sum^n x_i} \, dF(p).$$

We now introduce two axioms describing an estimator $\hat{p}$:

E1. $\hat{p}$ is unbiased for p, that is,

$$(\forall F) \, E_F \hat{p} = p = \int_0^1 x \, dF(x).$$

E2. For some loss function $\phi(\hat{p}, p)$ with ϕ convex in $\hat{p}$ for each p, $E\phi(\hat{p}, p)$ is a minimum uniformly in p; for example, $\phi(\hat{p}, p) = (\hat{p} - p)^2$.

From the Rao–Blackwell theorem [16, p. 121] and E2 we know that we can restrict our attention to $\hat{p}$ which are functions of the sufficient statistic $\sum^n x_i$, $\hat{p} = \hat{p}(\sum^n x_i)$. We now show that the relative-frequency is the unique unbiased function of $\sum^n x_i$. Let

$$\hat{p}(s) = \frac{s}{n} + \delta(s).$$

Noting that

$$P\left(\sum^n x_i = s\right) = \binom{n}{s} \int_0^1 p^s(1 - p)^{n-s} \, dF(p),$$

we have from E1 that

$$(\forall F) \sum_{s=0}^n \delta(s) \binom{n}{s} \int_0^1 p^s(1 - p)^{n-s} \, dF(p) = 0.$$

If we examine the special cases where F is a unit step at p and let $q = p/(1 - p)$, we find

$$(\forall q \geqslant 0) \sum_{s=0}^n \delta(s) \binom{n}{s} q^s = 0.$$

Invoking the unicity of polynomial representations we see that $(\forall s)\ \delta(s) = 0$. Hence $(1/n) \sum^n x_i$ is the unique unbiased function of the sufficient statistic and satisfies E1 and E2. We conclude that from the viewpoint of statistical decision theory finite relative-frequency can be reasonably justified as an estimator of probability when the order of performance of the experiments is irrelevant.

4. Bayesian Approaches

Bayes procedures are those of the third class, where it is assumed that there is some prior distribution for the unknown probability p. If we can supply a (nonempirical) prior, then we can calculate, by means of Bayes' theorem, a posterior distribution, based on the data and the probabilistic model for the data, for the unknown probability. The difficulty not heretofore encountered is the assessment of the prior distribution. Once we have the prior distribution we can invoke a variety of estimation procedures (e.g., maximum *a posteriori* estimation, minimum expected loss for some loss function) to obtain a numerical estimate of p.

There are both objective and subjective procedures for obtaining a prior. Under objective procedures we find classical and logical probability and other uses of invariance arguments to formalize a balance of information about the location of p. Subjective and personalistic procedures are those in which an individual, possibly after some more or less formal introspection, announces a prior distribution incorporating whatever knowledge he possesses. We postpone discussion of classical probability to Chapter VI, logical to Chapter VII, and subjective to Chapter VIII.

The illustration we give of invariance arguments is due to Laplace. If we have "no" prior knowledge concerning the location of p, then it seems reasonable to claim that the prior F is such that it is as probable that p is in the interval $(a, a + \epsilon)$ as that it is in the interval $(b, b + \epsilon)$ also of length ϵ; of course we assume that $a, b \geqslant 0$ and $a + \epsilon, b + \epsilon \leqslant 1$. If we disagree with this statement, then it seems that we are claiming to know that p is more likely to be in, say, $(a, a + \epsilon)$ than in $(b, b + \epsilon)$; this would be knowledge of the location of p, which we have disavowed having. The formal axiom then is as follows.

For all $a, b, \epsilon \geqslant 0$ such that $a + \epsilon, b + \epsilon \leqslant 1$,

$$F(a + \epsilon) - F(a) = F(b + \epsilon) - F(b).$$

Noting that since $p \in [0, 1]$, $F(0) = 0$, and setting $b = 0$, yields

$$F(a + \epsilon) = F(a) + F(\epsilon) \qquad (a, \epsilon \geqslant 0, a + \epsilon \leqslant 1).$$

This is the well-known Cauchy equation. Invoking the fact that distribution functions are bounded enables us to assert that all solutions to the functional equation are of the form

$$F(a) = Ca.$$

The constant C is established by the condition $F(1) = 1$. We conclude that the unique prior is the uniform distribution

$$F(a) = \begin{cases} 1 & \text{if} \quad a \geqslant 1 \\ a & \text{if} \quad 0 \leqslant a \leqslant 1 \\ 0 & \text{if} \quad a < 0. \end{cases}$$

Criticism of this conclusion and discussion of other objective arguments to establish priors are deferred to Chapter VI. While the use of invariance arguments to formalize some prior information is well established in statistics, it seems unreasonable and fraught with danger when applied to the formalization of an absence of prior information.

5. What Has Been Accomplished

The axiomatic approach "solves" the problem of measuring probability from data. However, how are we to use such probabilities? What predictions about experimental outcomes are we justified in making from knowledge of axiomatic probability estimates? No matter how we derive finite relative-frequency as a good estimator of probability, it is still heir to the ills of finite relative-frequency discussed in Section IVB. A possible justification for axiomatic estimators might be provided by the pragmatic vindication of Reichenbach, which is discussed a little more fully in Section IVK. Perhaps, following Reichenbach, we may maintain that, while unable to guarantee the success of extrapolations based on estimated probabilities, at least we are acting in a manner that can achieve success if success is achievable by any rational procedure. Nevertheless, the results obtained by any of the methods we have outlined do not inspire confident application.

IVK. Measurement of Comparative Probability: Induction by Enumeration

1. Introduction

While we are sympathetic to the attempts sketched in the preceding section to overcome the deficiencies of a limit of a relative-frequency interpretation of probability satisfying the Kolmogorov axioms, it seems obvious that there are still significant obstacles remaining to the achievement of a theory of probability having value for practice. More or less reasonable, but still somewhat arbitrary, principles have been adduced to enable us to measure probability from data of a relative-frequency sort. However, how are we to interpret or use these estimated probabilities or even the "true" probabilities? Why do we go to the trouble (e.g., much of the field of statistics) to estimate quantitative probability when, as we discussed in Chapter III, there seems to be little rationale behind the specificity of the probability concept?

The many difficulties that beset relative-frequency and probability in their relation to practice suggest that the attempt to estimate and use probability may be an ill-founded venture. We may be striving for more than is possible. Not only must we give up a deterministic conception of many phenomena but possibly also a probabilistic one. Perhaps all that we can assess is the comparative probability of occurrence of each outcome. Rather than attempt to relate quantitative probability to data (experience), we attempt the easier task of relating comparative probability to data. The measurement of a comparative (order) probability relation could presumably be done through the process of induction by enumeration.

2. Defining Induction by Enumeration

Induction by enumeration is the apparently obvious and explicit procedure of determining which of a set of events is most likely to occur by comparison of the past frequencies of occurrence of the events —that event which occurred most often in the past, under circumstances similar to the present one, is the one we predict to occur most often in the future. For example if $k > n/2$ heads are observed in n tosses of a coin, then a head is at least as probable as a tail in a future toss of the same coin. Seemingly this reduces inductive reasoning, an important special case of which is probability theory, based on data (there are non-data-based forms) to its skeletal and most compelling form. If we reject induction by enumeration, then what meaning can we assign to prediction from experience?

Interestingly enough the seemingly straightforward estimation of a probability order relation through induction by enumeration suffers from significant ambiguity. The use of enumerative induction requires two things: specification of the relevant set of past observations—the reference class of events; and an ability to recognize what is a confirming instance of the occurrence of an event in the data.

In practice the choice of a suitable reference class is neither easily formalized nor easily made. In the coin-tossing example should we base our estimate of the comparative probability of occurrence of heads and tails on the few trials recently made with the particular coin, or should we include the much more extensive data available on seemingly similar coins and on coin tossing? What do we do in the case of never before performed experiments such as the reliability of a unique system? In assessing the comparative probabilities of the potential causes of an individual's death, should we use the extensive data for the total population of the United States, or the much more limited data available on other individuals closely matched to the subject; if we insist on too close a match, then we are reduced to individuals sharing the significant characteristic that they are not yet dead, and this provides us with no data. In general the estimate of the comparative probability relation for a set of events will depend on the choice of reference class. However, beyond the vague and conflicting rules that we want as much data as possible and we want the data to come from experiments as similar as possible to the one whose outcomes we are predicting, we have no formal and unambiguous guidelines for the choice of a reference class.

A second and subtle source of ambiguity in the definition of enumerative induction is the nature of a confirming instance for the occurrence of an event. This question has been extensively studied under the heading of the logic of confirmation [17]. We will not delve deeply into these discussions by philosophers of science and induction. However, two classical examples of the difficulty may suffice to expose the problem. The first example is that the observation of "a black crow" is a confirming instance of the hypothesis that "all swans are white" (or the event that the next swan we observe will be white). The statement "a swan is a bird and all swans are white" is logically equivalent to the statement "all nonwhite birds are nonswans." Observation of a black crow clearly confirms the second statement and, therefore, also the first statement. Nevertheless, we would be reluctant in practice to feel that observation of a black crow makes it more probable that the next swan we see will be white rather than black.

The second example is the (in)famous Goodman [18] paradox leading to counterinductive policies. Let us assume that in n past observations on swans (made before 1980) all were white. Does this justify asserting that a swan observed after 1980 is more probably white than black? Let us introduce the new predicates "whack" and "blite," where "whack" means white prior to 1980 and black after 1980 and "blite" means black prior to 1980 and white after 1980. Any description involving black or white could as well be given using blite and whack, however strained the conversion might seem. On logical and linguistic grounds it is difficult to eliminate predicates like "whack" and "blite" from consideration. We now note that our present data on swans are that all were whack. Hence, we would be encouraged to predict that swans after 1980 will be black! However unfair this argument may seem at first glance, it has proved to be quite difficult to dispose of. Thus the estimation of comparative probability from frequency data through induction by enumeration is seen to be fraught with ambiguity. If these ambiguities could be removed to reveal something like our naive concept, of what value would it be?

3. Justifying Induction by Enumeration

How can we justify the apparently rational procedure of extrapolating what has occurred most frequently in the past as that which will occur most frequently in the future? The fact that this represents a strongly held belief, common even with animals, is no reason to continue its acceptance. In the 200 years since Hume the justification of induction has proved to be one of the most perplexing problems of philosophy. In addition to Hume's classic exposition [19] we can refer to modern discussions by Goodman [18], Black [20], Carnap [21], Feigl [22], Katz [23], von Wright [24], and many others. In this subsection we only superficially sketch a few of the arguments that have been advanced for and against induction.

Hume seems to have been the first to argue cogently the need for a justification of induction and the inadequacy of the unsophisticated justification that "it works." The argument from experience that enumerative induction, in predicting for the future what occurred most often under similar circumstances in the past, "works" is a circular one. Presumably by saying "it works" we mean that an inspection of n instances in the past where induction was used reveals that it provided the correct extrapolation more frequently than other "reasonable" procedures we can think of. Hence from the past success of induction, if indeed it really has been successful in the past, we are led to rely upon its future success. But this is just the inductive argument itself applied to the justification of induction.

One way in which this vicious circle can be broken is by asserting a metaphysical principle that there is uniformity or regularity in nature —the occurrence of phenomena exhibit long run regularities. Such foundation principles are not easily applied. Clearly nature is neither constant nor uniform in all respects. What distinguishes those variable phenomena for which there is regularity from those for which there is not? Even if we accept the existence of uniformities, we may still be unable to recognize them.

Furthermore, the principle of the existence of uniformities in nature does not assert that the next observation will necessarily be that which occurred most often in the past, but merely that in the long run what occurred most often in the past will occur most often in the future. Of what value are assurances of success in the exceedingly ill-defined "long run." What have we accomplished in, say, the design of a system for which the frequency of future successful operations will be high if we are willing to wait an indefinitely long time (the system only has a finite lifetime) and if our design data were typical of the long run of data? Most problems are not known to satisfy all of these conditions. Finally, to know what is a long run is to know more than a single number of trials; 10,000 tosses of a coin can also be thought of as a single toss of a 2^{10^4}-sided object.

We seem to be unable to justify enumerative induction as necessarily leading to many correct forecasts. Perhaps the root of the difficulty lies in our attempting an impossible and unnecessary justification. A justification for enumerative induction, and hence for its use in estimating comparative probability, on pragmatic rather than epistemic (leading to truth) grounds has been proposed by Reichenbach [25]. The position taken by Reichenbach on the problem of induction from repeated trials was that while we could not refute the skeptical thesis of Hume that no guarantee of success through induction was possible, this thesis was not critical. We could not validate induction, based on enumerating the past outcomes and selecting the most frequently occurring past outcome as our induction to the as yet unobserved trials, as leading to the "truth," but we could vindicate it as a pragmatic method. We could shift from the search for knowledge to the search for decision or behavior rules. If there is no regularity "binding" successive observations, then no method we could arrive at for induction could be shown to be preferable to any other method; in such a world success or failure is independent of our efforts. However, if there is regularity in the sequence of outcomes, then by following a rule based on extrapolating the outcome most frequently observed in the past we will be correct more often than by not doing so. We cannot guarantee that there is regularity in the observations, as pointed out by Hume, but we can act in such a manner as to achieve success as "frequently as possible" if success is achievable at all. This dictum is the central theme of Reichenbach's position and is used by him to defend the selection of relative-frequency as the guide to action.

A restatement of Reichenbach's position is offered by Feigl[†] [22, p. 130].

> The method of induction is *the only one for which it can be proved* (deductively!) that it leads to successful predictions *if* there is an order of nature, i.e., if at least some sequences of frequencies do converge in a manner not too difficult to ascertain for human beings ... .

While Reichenbach's justification is an improvement and opens the door to questions about the meaning of a justification, it does not seem to us to be satisfactory. Even if we ignore the ambiguities of enumerative induction, there are in fact other procedures asymptotically equivalent to enumerative induction which lead to conflicting predictions for any finite amount of data. Why should we favor enumerative induction over these other procedures? Reichenbach attempted to argue through the desirability of "descriptive simplicity," but this seems to be very weak if not irrelevant grounds upon which to prefer enumerative induction. Furthermore, how pragmatic is this pragmatic defense? If we cannot achieve some indication of success, then perhaps it is best not to expend our efforts on the problem. In practice we look with disfavor upon unicorn hunts, even those designed to catch a unicorn if any hunt can.

IVL. Conclusion

We have found that even the easier task of estimating comparative probability from relative-frequency data harbors unsuspected difficulties. The unresolved aspects of comparative probability, when combined with the discussions of earlier sections, lead us to doubt that there is any value in a concept of quantitative probability based *solely* on relative-frequency. This view seems to have been implicitly shared by Born,

---

[†] H. Feigl, De Principiis Non Disputandum ... ?, in *Philosophical Analysis* (M. Black, ed.), p. 130. Englewood Cliffs, New Jersey: Prentice Hall, 1963.

the physicist who first interpreted the modulus of the Schroedinger wave function as probability. To quote Born[†]

> Another *metaphysical* principle is incorporated in the notion of probability. It is the *belief* that the predictions of statistical calculations are more than an exercise of the brain, that they can be trusted in the real world (italics added) [26, p. 124].

What the justification of this belief is to be, which statistical calculations are intended, and what is meant by "trust" are apparently unexplained by Born.

It seems that we must look elsewhere for a useful interpretation of uncertainty, chance, and probability. The first direction we examine developed from work of von Mises on the relative-frequency approach.

References

1. B. de Finetti, Foresight: Its Logical Laws, Its Subjective Sources, in *Studies in Subjective Probability* (H. Kyburg and H. Smokler, eds.), pp. 95–158. New York: Wiley, 1964.
2. M. Loeve, *Probability Theory*, p. 400. Princeton, New Jersey: Van Nostrand-Reinhold, 1960.
3. A. Birnbaum, Is There a Basic Role for Frequency or Subjective Interpretations of Probabilities in the Natural Sciences? Discussion draft of February 1969. Courant Inst., New York Univ., New York.
4. B. Gnedenko, *Theory of Probability*, 4th ed. Bronx, New York: Chelsea, 1968.
5. H. Cramér, *Mathematical Methods of Statistics*. Princeton, New Jersey: Princeton Univ. Press, 1957.
6. M. Fisz, *Probability Theory and Mathematical Statistics*, 3rd ed. New York: Wiley, 1963.
7. E. Parzen, *Modern Probability Theory and Its Applications*. New York: Wiley, 1960.
8. T. Fine, On the Apparent Convergence of Relative Frequency and Its Implications, *IEEE Trans. Information Theory* IT-16, 251–257, 1970.
9. A. Kolmogorov, On Tables of Random Numbers, *Sankhyā Ser. A*, p. 369, 1963.
10. A. Church, On the Concept of a Random Sequence, *Bull. Amer. Math. Soc.* **46**, 130–135, 1940.
11. R. von Mises and H. Geiringer, *The Mathematical Theory of Probability and Statistics*, pp. 7–43. New York: Academic Press, 1964.
12. J. Ville, *Étude Critique de la Notion de Collectif*, pp. 58–63. Paris: Gauthiers-Villars, 1939.
13. A. Kolmogorov, Logical Basis for Information Theory and Probability Theory. *IEEE Trans. Information Theory* 14, 663, 1968.
14. I. J. Good, *Estimation of Probabilities*. Cambridge, Massachusetts: Technology Press, 1965.
15. W. C. Salmon, *The Foundations of Scientific Inference*, pp. 96–102. Pittsburgh, Pennsylvania: Univ. of Pittsburgh Press, 1966.

[†] M. Born, *Natural Philosophy of Cause and Chance*, p. 124. New York: Dover, 1964.

16. T. Ferguson, *Mathematical Statistics*. New York: Academic Press, 1967.
17. C. G. Hempel, *Aspects of Scientific Explanation*. New York: Free Press, 1965.
18. N. Goodman, *Fact, Fiction and Forecast*, pp. 75–80. Cambridge, Massachusetts: Harvard Univ. Press, 1955.
19. D. Hume, Skeptical Doubts Concerning the Operations of the Understanding, *David Hume on Human Nature and Understanding*, Pt. II. New York: Collier Books, 1962.
20. M. Black, *Margins of Precision*, pp. 57–178. Ithaca, New York: Cornell Univ. Press, 1920.
21. R. Carnap, *Logical Foundations of Probability*, 2nd ed. Chicago, Illinois: Univ. of Chicago Press, 1962.
22. H. Feigl, De Principiis Non Disputandum ... ?, in *Philosophical Analysis* (M. Black, ed.), pp. 113–147. Englewood Cliffs, New Jersey: Prentice-Hall, 1963.
23. J.Katz, *The Problem of Induction and Its Solution*. Chicago, Illinois: Univ. of Chicago Press, 1962.
24. G. von Wright, *The Logical Problem of Induction*, 2nd ed. New York: Barnes & Noble, 1957.
25. H. Reichenbach, *The Theory of Probability*, p. 475. Berkeley: Univ. of California Press, 1949.
26. M. Born, *Natural Philosophy of Cause and Chance*. New York: Dover, 1964.

V

Computational Complexity, Random Sequences, and Probability

VA. Introduction

Computational complexity has recently been proposed as a basis for the recognition or definition of random sequences and the definition or measurement of probability. To some degree this approach is an outgrowth of von Mises' idea of a collective and the role of place-selection functions. While von Mises discussed infinite sequences, our chief interest is with finite sequences—the only ones encountered in practice. Kolmogorov reformulated the use of place-selection functions so that they could be applied to finite sequences. There remained the problem of the choice of an appropriate family of such functions. Reflection on this question leads us from reliance on a pragmatic impossibility-of-a-gambling-scheme view of random sequences to an epistemic view that a random sequence is highly irregular or complex. We then examine a form of Kolmogorov's approach to the complexity of a sequence through the shortest programs needed to compute the sequence. After a brief discussion of statistical tests, we relate conditional complexity to statistical critical level. We are then able to provide complexity-based definitions of random and exchangeable finite se-

118

quences and independent exchangeable sequences. The problem of random infinite sequences is also briefly discussed.

After a brief discussion of the application of complexity to the measurement of comparative probability, we follow Solomonoff and present two of his definitions of quantitative probability. It is evident that much remains to be done to achieve and justify an adequate computational-complexity basis for probability.

VB. Definition of Random Finite Sequence Using Place-Selection Functions

Kolmogorov [1] suggested that a given (binary) sequence S_N of length N ($\geqslant n$) be called (n, ϵ) random with respect to a finite family Φ of admissible algorithms $\{\phi_i\}$ (generalized place-selection rules in that selection of x_i can depend on x_j for $j > i$) if there is some p such that the relative-frequencies of 1's in the subsequences S^i of length at least n generated by applying ϕ_i to S_N (some ϕ_i will be discarded as yielding S^i that are too short) all lie within ϵ of p; that is, the relative-frequency in this finite sequence is approximately (to within ϵ) invariant under any of the methods of subsequence selection which yield subsequences of length at least n. Kolmogorov has shown that if the number of admissible algorithms does not exceed $\frac{1}{2} \exp[2n\epsilon^2(1 - \epsilon)]$, then for any p, and any $N \geqslant n$, there is some sequence S_N that is (n, ϵ) random with respect to that family of place-selection rules. This definition of degree of randomness closely adheres to the von Mises suggestion of a collective for infinite sequences. However, what are the guidelines for the choice of Φ so that the resulting judgments agree with our informal notion of a random sequence?

The family of admissible algorithms or place-selection rules Φ tests for regularities (e.g., properties that could be used to a gambler's advantage) in the candidate sequence. For example, if we are interested in testing the finite sequence $S_n = x_1, ..., x_n$ for "periodicity" (i.e., regular spacing of 1's), then we might select the family of place-selection functions $\Phi = \{\phi_i\}$ where

$$\phi_i(S_k) = \begin{cases} 1 & \text{if} \quad i \geqslant k \\ x_{k-i} & \text{if} \quad k > i \end{cases} \quad i = 0, 1, ...$$

This simple form of definition can be rewritten with integer encoding as discussed in Section IVE. If S_n has 1's spaced i terms apart, then application of $\phi_i \in \Phi$ would yield a subsequence with initial segment S_i followed by 1's. For sufficiently large n the relative-frequency of 1's would approach 1 for the subsequence but could be as low as $1/i$ for the original sequence. Thus the difference between the relative-frequency in S_n and in the subsequence generated by ϕ_i would cause us to, quite reasonably, reject S_n as being random. If we take a sufficiently large collection Φ and if there is a simple pattern in the candidate sequence S_n of length n, then presumably some $\phi \in \Phi$ will detect it by selecting a subsequence having properties (relative-frequency) differing from S_n. The basic question then is what constitutes a proper choice of Φ? Too keen an eye for pattern will find it anywhere. For infinite sequences, the countable collection of effectively computable ϕ seemed reasonable. For finite sequences we must be more restrictive. For example, there are only 2^n distinct binary sequences of length n, and for each such sequence the effectively computable place-selection function $\phi(S_{i-1}) = x_i \in S_n$ would yield a subsequence of all 1's; there would be no nontrivial random sequences.

Informally we would expect to define a random finite sequence by including in Φ only simple, straightforward, easily described functions and not complicated or contrived functions which could only be described by a table of values [e.g., $\phi(S_{i-1}) = x_i$ above]. Focusing on the candidate sequence rather than on Φ, it appears that we are requiring a random sequence to be very irregular; only in this way could it fail the place-selection test for regularity. Instead of discussing what a simple place-selection function might be, perhaps we should discuss the problem of what constitutes a complex (random) sequence? A review of further work related to place-selection approaches to the definition of random sequences is provided by Knuth [2] and Martin-Löf [3].

A random binary sequence should correspond to a complicated function from the integers to $\{0, 1\}$, not to a simple function. Alternatively, the generating mechanism for a random sequence should not be describable by a simple law. However, what beyond a cultural accident distinguishes a simple from a complex function or sequence? Following Kolmogorov and Martin-Löf, we propose a definition of complexity in the next section and then derive some of the consequences of the definition. This definition agrees with one suggested in Section IVC.

VC. Definition of the Complexity of Finite Sequences

1. Background to Absolute Complexity

It was noted by Chaitin, Kolmogorov, and Solomonoff that a classification of the computational complexity of sequences was available from the theory of computability and automata. Our first step is to present such a classification of sequences and then to explore its consequences. In particular, we will show that a definition of randomness can be based on complexity that will imply any statistical (axiomatic probability and independent and identically distributed trials) definition of randomness.

By the complexity of a sequence we mean the difficulty in describing the sequence or in storing the sequence through some description. A description of the sequence is a set of instructions for some computer that enables it to reconstruct the sequence. Hence we need both a notion of the difficulty or awkwardness of a description and a suitable class of computers. For simplicity in the ensuing discussion we will assume that the elements x_i of the sequence S_n of length n can take on only the values 0 and 1. The procedure for computing S_n can be thought of as a computer, machine, or algorithm α which accepts as input a program (description) P and generates as output a sequence S_n, $S_n = \alpha(P)$.

A program is a finite binary sequence. The algorithm, machine, or computer α is a mapping from a subset of the space of finite binary sequences into the space of all binary sequences; we have more to say about reasonable classes of computers in the sequel. For example, we might enumerate all of the FORTRAN symbols and instructions (DO, IF, etc.) by binary numbers and thereby rewrite a FORTRAN program listing as a binary string P; or P might be the object deck (machine language) that results from compiling a FORTRAN program. What then determines the "awkwardness" of a description P of S_n or the difficulty of storing P?

If we are simply interested in a hierarchy of complexities of description, then we might be satisfied by an order relation on the set of binary sequences. This order relation might be represented numerically for convenience. However, such a representation would not be unique; any increasing function of a representation is also a representation of the same ordering. If we are interested in difficulty of storage, then a numerical representation may be less arbitrary. The smallest number of bits needed to store a program for a sequence on a given computer provides an absolute scale for the measurement of difficulty of storage. While it is not clear that we need to restrict ourselves to complexity as a measure of the difficulty of storage, we will assume that a numerical representation of complexity is desired. By a measure of complexity we will mean a mapping from the set of binary sequences to the integers which at least satisfies adequacy criteria to be proposed below.

One obvious measure of storage difficulty is the length $l(P)$ of P; each bit of P requires a bit of memory. However, P is itself a binary sequence and perhaps there is some simpler description P' of P such that $l(P') \ll l(P)$. Hence, whatever particular measure $K_\alpha(S)$ we select to measure the complexity of S relative to α we impose the following criterion of adequacy:

A1.
$$K_\alpha(S) \approx \min\{k : \alpha(P) = S, k = K_\alpha(P)\}.$$

The approximate equality is in place of the strict equality that might not be achievable on a given computer α and also to allow for a small discrepancy in complexity that might be required to distinguish between outputs and inputs for α. In general there will be many programs for a given sequence S. The complexity of S should be judged from that of the least complex way of storing it, and hence, we take a minimum over all programs for S.

The particular measure of complexity chosen by Kolmogorov [4] and the one for which the complexity-based approach has been developed is
$$K_\alpha(S) = \min\{l : \alpha(P) = S, l = l(P)\},$$

where it is understood that if there is no P such that $\alpha(P) = S$, then $K_\alpha(S) = \infty$. Applying A1 we see that the Kolmogorov definition is inadequate unless

$$\min\{l : \alpha(P) = S, l = l(P)\}$$
$$\approx \min\{k : \alpha(P) = S, k = \min\{l : \alpha(P') = P, l = l(P')\}\}$$
$$= \min\{k : \alpha(\alpha(P')) = S, k = l(P')\}.$$

This condition constrains the choice of computer α that can be reasonably used to implement the Kolmogorov measure.

Other criteria of adequacy that further constrain the choice of complexity measure and, through the Kolmogorov definition, the choice of computer are

A2. $K_\alpha(S)$ should be finite for all finite-length sequences.

A3. There are only finitely many sequences having the same finite complexity.

A4. Complexity evaluations should be conservative, at least in the sense that if K_α is our selected measure and K' any other "reasonable" measure satisfying A1–A3, then

$$(\exists c)(\forall S)\, K_\alpha(S) < c + K'(S).$$

A5. The notion of a complex sequence should be compatible with the statistical notion of a random sequence.

The motivation for A2 is that no finite sequence can be infinitely irregular or difficult to store; at worst we can construct a table for the sequence that requires finitely many bits of memory. Criterion A3 is intended to rule out trivial complexity measures such as the constant function. It is suggested by work of Pager [5].

In evaluating the complexity of a sequence we want to be sure to consider the most efficient descriptions. Hence, as suggested by A4, an adequate measure of complexity should not yield values that, depending on the sequence, can be arbitrarily larger than those assigned by any other "reasonable" measure of complexity. If we do not define "reasonable" beyond satisfying A1–A3, then there does not exist any measure adequate in the sense of A4. However, there will be adequate measures if we restrict ourselves to complexity measures of the form suggested by Kolmogorov.

Finally, we desire a useful notion of complexity. Our purposes concern the development of a theory of random phenomena, and not, say, a theory of speed of computation. Criterion A5 makes this point, although it is neither very specific nor exhaustive of our interests in introducing considerations of complexity.

2. A Definition of Absolute Complexity

The adequacy criteria A1–A5 do not lead us to a unique measure of complexity. For definiteness, we elect to implement Kolmogorov's definition of complexity in terms of minimum program length by restricting α to a subclass of universal Turing machines (UTM), as was done in Section IVC. After a brief description of the class of computers we adopt, we will proceed to verify the adequacy of our choice. However, we must emphasize that other choices of complexity measure and computers are possible and at least *a priori* appear to be reasonable.

A Turing machine (TM) may be thought of as a machine with a set of internal states, a tape partitioned into tape cells containing a single entry from a finite alphabet (including the blank symbol), a read/write head that can read the symbol in a tape cell and write a symbol in that cell (including erasing the present symbol or leaving it unchanged), and an ability to move the read/write head either left or right one cell at a time. It is a discrete time system. Somewhat more formally a TM is a finite collection of quintuples $\{(S, R, W, M, S')\}$ where S is the present internal state, R is the symbol read on the tape cell, W is the symbol written (overprinted) on the same tape cell in place of R, M is a tape-movement instruction of left, right, or stay, and S' is a new internal state of the TM. Equivalently, we may combine W and M into an act A and consider a TM as a finite set of quadruples $\{(S, R, A, S')\}$. The only constraint on the definition of a TM is that the collection of quintuples be consistent; that is, there do not exist two quintuples $(S_i, R_i, W_i, M_i, S_i')$, $(S_j, R_j, W_j, M_j, S_j')$ in the set such that $S_i = S_j$, $R_i = R_j$ yet $(W_i, M_i, S_i') \neq (W_j, M_j, S_j')$. For convenience we take the tape alphabet to be ternary $\{0, 1, \text{blank}\}$. The program is a binary $\{0, 1\}$ string entered onto the tape. Computation starts with some convention such as the TM in state S_1 and the read/write head at the left end of the program sequence. The solution to the computation is the binary sequence on the tape after the machine has gone into a "halt state" signaling the termination of the calculation; a halt state is reached when W, M, and S' are such that at the next operation there is no quintuple corresponding to $S = S'$ and the symbol being read.

All Turing machines can be effectively enumerated in that each finite collection of quintuples can be assigned an index number (by means of some "reasonable" algorithm) and given that index number (e.g., Gödel number) we can effectively determine the corresponding machine; there are many ways to carry out the enumeration. A universal Turing machine (UTM) is a finite-state TM that can simulate any other TM and thereby calculate any partial recursive function. The UTM may be visualized as accepting a program tape with two inputs, the first the index number of the machine it is to simulate and the second the input to the simulated machine.

To be more precise, we effectively enumerate all consistent, finite sets of quintuples and thereby assign an index to each TM (partial recursive function); $\varphi_x(P)$ is the output sequence of the xth TM with program input P, when this output is defined. For many TM there exist programs such that the halt state is never entered and computation continues indefinitely. In such cases the function $\varphi_x(P)$ is undefined.

To construct a UTM, we introduce the recursive encoding function $f_2(X, P)$ mapping pairs of finite binary sequences into a single finite binary sequence and having a unique inverse. Somewhere in our list of TM there is a machine φ_y which when given the program $f_2(X, P)$ performs the inverse to recover X, looks up φ_x in the list of TM, and calculates $\varphi_x(P)$. Hence $\varphi_y(f_2(X, P)) = \varphi_x(P)$ where $\varphi_x(P)$ is defined and is undefined otherwise. The machine φ_y is a TM that can simulate any other TM and is called a UTM. There are infinitely many UTM, each one capable of simulating the others.

The preceding discussion easily extends to cover the case of functions of several variables. By an encoding function f_k we mean a uniquely invertible, recursive mapping of k finite binary sequences into a single finite binary sequence; Ψ is a k-input UTM if it can be represented in terms of a 1-input UTM φ through

$$\Psi(X_1, ..., X_k) = \varphi(f_k(X_1, ..., X_k)),$$

and Ψ is universal in the sense that if we consider partial recursive functions θ of k-inputs, then

$$(\exists f_2)(\forall \theta)(\exists X)(\forall X_1, ..., X_k)(\theta(X_1, ..., X_k)$$
$$= \Psi(X_1, ..., X_{k-1}, f_2(X, X_k)) \text{ where defined}).$$

Further discussion of the concept of a UTM can be found in Rogers [6].

The subclass of UTM that is adequate for our purposes will be those Ψ for which the encoding function f_2 satisfies

$$(\forall X)(\exists c)(\forall P)\, l(f_2(X, P)) < c + l(P);$$

such Ψ will be called additive UTM (AUTM). We now propose

Definition 1. A sequence S has complexity $K_\Psi(S)$ if Ψ is an AUTM and

$$K_\Psi(S) = \min\{l : \Psi(P) = S, l(P) = l\}.$$

3. Properties of the Definition of Absolute Complexity

Central to the verification of the adequacy of the preceding definition is

Lemma 1. If Ψ is an AUTM, ϕ any TM, then

$$(\exists c)(\forall S)\, K_\Psi(S) < c + K_\phi(S),$$

where

$$K_\phi(S) = \min\{l : \phi(P) = S, l(P) = l\}.$$

Proof. All proofs in this chapter are contained in the Appendix to this chapter. ∎

As a simple consequence of Lemma 1 we have the following.

Corollary. If Ψ and ϕ are both AUTM, then

$$(\exists c)(\forall S) \mid K_\Psi(S) - K_\phi(S) \mid < c.$$

The adequacy of our definition of complexity is partially defended by

Theorem 1. K_Ψ satisfies the adequacy criteria A1–A3.

There does not exist any complexity measure satisfying A4 with respect to all measures satisfying A1–A3. If we restrict ourselves to the subclass of measures given by

$$\min\{k : \phi(P) = S, k = l(P)\},$$

where ϕ is any partial recursive function, then Lemma 1 verifies that measures based on any AUTM Ψ satisfy A4 with respect to this subclass. This is the motivation for our interest in AUTM-based complexity measures.

A sequence that is judged complex according to Ψ will be judged complex according to many other computers ϕ. Of course, there is the constant c in the bound, and c depends on ϕ although not on the sequence whose complexity is being evaluated. It is not possible to bound c over the class of all ϕ; this follows from consideration of special function generators—$(\forall S)(\exists \phi)$ $K_\phi(S) = 1$. Therefore, for any given sequence S judged complex by Ψ there will exist some special machine ϕ according to which S is not complex. However, from the corollary to Lemma 1, given any two AUTM, Ψ and ϕ, the resulting complexity evaluations must be approximately equal, at least for sequences that are complex according to either Ψ or ϕ. Hence, if S is judged as complex according to Ψ, then it should also be judged as complex according to ϕ, at least for very complex S. Thus judgments of complexity made according to an AUTM are "invariant" for fixed machines and sufficiently complex sequences but are reversible for a fixed sequence, however complex, by a proper choice of competing AUTM.

4. Other Definitions of Absolute Complexity

Are there grounds for further restriction of the class of machines suitable for complexity evaluation? Kolmogorov's feeling on this question is perhaps well expressed by [4, p. 6]:

> One must, however, suppose that the different 'reasonable' variants presented here will lead to 'complexity estimates' that will converge on hundreds of bits instead of tens of thousands. Hence, such quantities as the 'complexity' of the text of *War and Peace* can be assumed to be defined with what amounts to uniqueness.

However, Kolmogorov does not explain what he means by a "reasonable variant" of a computer.

Loveland has proposed a modification of the complexity definition which may be a step in the direction of selecting a "reasonable variant." Let S_j denote the initial segment of length j of S, Ψ be a 2-input AUTM, and J be the binary representation of j. Loveland [7] advances

Definition 2. The uniform complexity $K_\Psi(S; l(S))$ of S relative to Ψ is given by

$$K_\Psi(S; l(S)) = \min\{k : (\exists P)\,(\forall j \leqslant l(S))(\Psi(J, P) = S_j), l(P) = k\}.$$

The class of descriptions for S is narrowed to those programs which when supplemented by an integer j can generate the initial segment of length j of S, as well as being able to generate S. We omit Loveland's discussion of the merits of his definition.

Chaitin [8], who independently developed the notion of a complexity-based definition of a random sequence, considered the use of a class of 3-tape machines: The program appears on the first tape and is read unidirectionally; the scratchwork is performed on the second tape; and the output is irreversibly written on the third tape. A general approach to the notion of the length of a program has also been presented by Pager [5]. Other notions of complexity are reviewed in Hartmanis and Hopcroft [9].

5. Conditional Complexity

We introduce a concept of conditional complexity to measure the residual complexity of a sequence S given information I about S.

In the cases of immediate interest to us I will be either the length $l(S)$ of S or $l(S)$ and the weight $w(S)$ of S where

$$w(S) = \sum_{i=1}^{l(S)} x_i$$

is the number of 1's in the binary sequence S. The (absolute) complexity of a sequence by itself does not tell us whether the sequence is random. In general, the longer the sequence, the more complex we can expect it to be. What is of interest is the relative complexity of a sequence in a suitable reference class.

The concept of conditional complexity that we will use in the sequel to determine the relative complexity of sequences belonging to a given class is given by

Definition 3. The conditional complexity $K_\Psi(S \mid I_1 , ..., I_k)$ of S given information represented as k binary strings, $I_1 , ..., I_k$, satisfies

$$K_\Psi(S \mid I_1 , ..., I_k) = \min\{l : \Psi(I_1 , ..., I_k , P) = S, l(P) = l\},$$

where Ψ is a $k + 1$-inputs AUTM.

In other words, we measure the conditional complexity by the length of the shortest program which when adjoined to the given information becomes a program for the desired sequence.

The following properties of the conditional-complexity measure are easily derived along the lines of the proof of Lemma 1 and the verifications of the adequacy criteria.

Lemma 2. If $K_\Psi(S \mid I_1 , ..., I_k)$ is a conditional-complexity measure, then:

(a)   If ϕ is any partial recursive $k + 1$-input computer, then

$$(\exists c)(\forall S, I_1 , ..., I_k) \, K_\Psi(S \mid I_1 , ..., I_k)$$
$$< c + \min\{l : \phi(I_1 , ..., I_k , P) = S, l = l(P)\}.$$

(b)   If ϕ is an AUTM, then

$$(\exists c)(\forall S, I_1 , ..., I_k) \, (\mid K_\Psi(S \mid I_1 , ..., I_k) - K_\phi(S \mid I_1 , ..., I_k)\mid < c).$$

(c)   Let $N(I_1 , ..., I_k)$ be the number of sequences for which $(I_1 , ..., I_k)$ is true. Then

$$(\exists c)(\forall S, I_1 , ..., I_k) \, (K_\Psi(S \mid I_1 , ..., I_k) < c + \log_2 N(I_1 , ..., I_k)).$$

The similarity in the properties of conditional and absolute complexity makes it easy to verify A1–A4 for conditional complexity. We omit the details. There remains the pragmatic question posed by A5 and its generalization to the possibility of computational-complexity-based analogs of statistical models.

6. Ineffectiveness of Complexity Calculations

None of the proposed definitions of descriptive complexity yields effectively computable measures of complexity. That is to say, there is no TM β such that

$$(\forall S, I)\, \beta(S, I) = K_{\Psi}(S \mid I).$$

The nonexistence of such a β is a corollary of the well-known un-decidability of the halting problem. It might even be argued that no reasonable measure of descriptive complexity should be effectively computable. If β is a TM representing a proposed effective complexity measure in that $\beta(S, I)$ is the conditional complexity of S given I, then

$$(\exists c)(\forall S, I)\, K_{\Psi}(S \mid I) < c + \beta(S, I).$$

Hence method β strongly violates our adequacy criterion A4. Put another way,

$$(\exists c)(\forall n)(\exists S, I)\, K_{\Psi}(S \mid I) < c + \log n, \qquad \beta(S, I) > n.$$

Hence there exist arbitrarily highly β-sense complex sequences that are of very low Kolmogorov complexity.

Accepting that K_{Ψ} is not an effectively computable function, we can still consider evaluating it for specific arguments. Chaitin [10] has shown that attempts to prove any statement of the form "$K_{\Psi}(S) = n$" within a formal system, whose axioms are described by a binary sequence, require that the length of description of the formal system be approximately n. Furthermore, if we interpret the number of steps in a proof carried out within the formal system as computation time on a suitable machine, then there is no recursive function $f(n)$ that can be an upper bound to the number of steps required to determine the complexity of all sequences of length n. In other words, for large n, it can be exceedingly difficult to determine whether a given sequence is of high complexity. In view of Chaitin's observations we can only ask for reasonable effectively computable approximations to K_{Ψ}.

An interesting attempt at approximating Kolmogorov complexity has been suggested by Ziv [11]. He partitions a sequence $S = x_1, ..., x_n$ into successive blocks $S_{kl+1}^{k(l+1)} = x_{kl+1} \cdots x_{kl+k}$ and considers a class of measures where the complexity of S is approximately the sum of the conditional complexities of the blocks $\{S_{kl+1}^{k(l+1)}\}$. For this class of measures he is able to develop complexity bounds that depend on relative-frequency-based entropies.

VD. Complexity and Statistics

1. Statistical Tests for Goodness-of-Fit

We are concerned only with the question of the acceptance or rejection of probabilistic models of, or hypotheses concerning, finite-length binary sequences. By a probabilistic hypothesis we mean a family $\{P_\theta, \theta \in \Theta\}$ of probability measures where P_θ is a measure on the set of all finite-length binary sequences. A statistical test [12] T for $\{P_\theta, \theta \in \Theta\}$ is a family $\{T_\delta\}$ where T_δ is a set of sequences for which we reject the hypothesis at level $2^{-\delta}$. To define a test in greater detail, we introduce

Definition 4. $T = \{T_\delta\}$ is a test for $\{P_\theta, \theta \in \Theta\}$ if:

(a) $T_\delta \subseteq \{S : l(S) < \infty\}$;

(b) $\delta > \delta' \Rightarrow T_\delta \subseteq T_{\delta'}$;

(c) $\max_{\theta \in \Theta} P_\theta(S) = 0 \Rightarrow S \in T_\delta$;

(d) $(\forall \theta \in \Theta)\, P_\theta(T_\delta) \leqslant 2^{-\delta}$;

(e) T is recursive.

Requirement (c) has us reasonably reject the probabilistic hypothesis if we observe a sequence S having probability 0 under the hypothesis. Requirement (d) asserts that if we base our decision on T_δ, then the probability of falsely rejecting the hypothesis is no more than $2^{-\delta}$. Presumably those sequences in a T_δ corresponding to small $2^{-\delta}$ are those for which we feel confident about rejecting the hypothesis. Hence, they should also be included in larger rejection regions $T_{\delta'}$ ($\delta' < \delta$), as required by (b).

The final requirement (e) asserts that for any S we should be able to (algorithmically) determine effectively whether it is, or is not, in each T_δ; that is, the test T can always be applied.[†] Each effective test

---

[†] We prefer (e) to the weaker requirement, proposed by Martin-Löf [13], that T only be recursively enumerable.

T can be described by a characteristic, or indicator, function $t(\delta, S)$ that is a recursive function satisfying

$$t(\delta, S) = \begin{cases} 1 & \text{if } S \in T_\delta \\ 0 & \text{if } S \notin T_\delta . \end{cases}$$

As there are only a countable number of recursive functions and computable numbers δ, there can only be a countable number of effectively computable statistical tests. We assume an effective enumeration $\{T^{(i)}\}$ of the set $\mathcal{T}$ of computable tests for a given hypothesis $\{P_\theta, \theta \in \Theta\}$.

The outcome of applying a test T to a sequence S can be summarized through the critical level $2^{-m_T(S)}$ where

$$m_T(S) = \sup\{\delta : S \in T_\delta\}.$$

The larger $m_T(S)$ is, the smaller is the critical level, and the less reasonable is it to accept the probabilistic hypothesis. Large $m_T(S)$ corresponds to rejection of the hypothesis by T with a low probability $[\leqslant 2^{-m_T(S)}]$ of error. Finally, since $\{T_\delta\}$ are nested, knowing m_T provides a complete description of the outcome of T. In view of the preceding properties of m_T we propose

Definition 5. T accepts $\{P_\theta, \theta \in \Theta\}$ on the basis of the observation S if for preselected threshold t,

$$m_T(S) < t.$$

2. Universal Statistical Tests

The notion of a universal test, introduced by Martin-Löf [13], is indirectly defined in

Definition 6. R is a universal test for $\{P_\theta, \theta \in \Theta\}$ if $R \in \mathcal{T}$ and

$$(\forall T \in \mathcal{T})(\exists c)(\forall S)\, m_T(S) < c + m_R(S).$$

A constructive description of a universal test, based on the enumeration $\{T^{(i)}\}$ of $\mathcal{T}$, is given by

Theorem 2. $R \in \mathcal{T}$ is a universal test only if

$$(\exists\{c_i\})(\forall R_\delta \in R)\, R_\delta \supseteq \bigcup_{i=1}^{\infty} \bigcup_{\delta' \geqslant \delta + c_i} T_{\delta'}^{(i)}.$$

The significance of a universal test R is that sequences of large critical level according to R (small m_R) also tend to be of large critical level according to other tests $T \in \mathcal{T}$.

While universal tests are of interest from the viewpoint of selecting good tests, our results will apply to all tests in $\mathcal{T}$, and there is no reason to restrict their statement to universal tests. The relationship we describe between complexity and statistical concepts will, however, be weaker than that stated by Martin-Löf. This discrepancy is attributable to Martin-Löf's inclusion [13, p. 607] of certain ineffectively computable tests (the rejection regions were determined by the ineffectively computable complexity of the sequences) in the universal tests.

3. Role of Complexity in Defining Probabilistic Models

A first approach to the definition or testing of certain probabilistic models or hypotheses by the use of conditional complexity is to identify an appropriate collection $\mathcal{I} = \{I_j\}$ of subsets I_j of the set of finite-length binary sequences, where $\bigcup_j I_j$ is exhaustive of all finite binary sequences, and then to adopt the rule of rejecting the hypothesis unless the observation S is of high conditional complexity given $S \in I \in \mathcal{I}$. Definitions of probabilistic models based on a criterion of high conditional complexity, relative to the information supplied by $\mathcal{I}$, only seem reasonable when we can find $\mathcal{I}$ such that

(a) $(\forall I \in \mathcal{I})(\forall S, S' \in I) \sup_\theta P_\theta(S \mid I) = \sup_\theta P_\theta(S' \mid I)$;

(b)   there are "few" $I \in \mathcal{I}$ of "small" cardinality;

(c)   each I is recursive.

If all sequences in each $I \in \mathcal{I}$ are equally probable, then the decision as to whether to reject the hypothesis upon observing S becomes one of whether S is random in the subset I. As we suggest in Section VE, high conditional complexity appears to be a good criterion for judgments of randomness. Furthermore, as judgments of complexity involve arbitrary constants and are only reasonable when applied to highly complex objects, we would need most $I \in \mathcal{I}$ to be large sets. If we cannot find an $\mathcal{I}$ that reduces the problem to a question of identifying random sequences, or if any such $\mathcal{I}$ is judged inappropriate, then it is not evident how maximal complexity can be applied to the definition of tests for probabilistic hypotheses.

A second approach concentrates on a complexity-based counterpart to statistical tests, rather than to probabilistic models, and the following is easily verified. Given any collection $\{T^\alpha\}$ of tests satisfying

(a) $\left(\bigcup_\alpha \bigcap_\delta T_\delta^\alpha\right)$ is recursive,

(b) $(\exists\{j_\alpha\})\left(\bigcap_\alpha \overline{T_{j_\alpha}^\alpha}\right)$ is recursive and infinite,

then we can always find a recursive partition $\mathscr{I} = \{I_j\}$ of the set of finite-length binary sequences such that

(a) For each n only finitely many of $\{I_j\}$ satisfy $\| I_j \| < n$,

(b) $(\exists c)(\forall S)\, K_\Psi(S \mid I) > c \Rightarrow (\forall\alpha)\, m_{T^\alpha}(S) \leqslant j_\alpha$.

Hence it is often possible to find a (perhaps unnatural) collection $\mathscr{I}$ such that if S is of high conditional complexity given $I \in \mathscr{I}$ then S also passes any test T^α at a critical level of at least 2^{-j_α}. In general such $\mathscr{I}$ are not unique.

A third alternative may be to accept the existence of countably many suitable $\mathscr{I}$ and complexity-based tests for a given probabilistic hypothesis and not attempt to select a unique test. Perhaps one could then develop a "universal" complexity-based test. Finally we need to know more about the relation between complexity-based tests and the complexity-based definitions of probability to be discussed in Section VJ. These issues require further research and we are unable to resolve them here.

In Sections VE–VH we will identify appropriate natural $\mathscr{I}$ and proceed as first outlined. As will be seen in Subsection VG2, this approach is of limited applicability.

4. A Relation between Complexity and Critical Level

Central to the validation of complexity-based definitions of probabilistic hypotheses of the first type discussed above is the relation between the rejection regions based on low conditional complexity $K_\Psi(S \mid I)$, where we abuse our notation by taking I to denote interchangeably a set of sequences and a description of the set of sequences, and the statistical rejection regions based on large m_T.

Theorem 3.

(a) $(\forall T \in \mathcal{T})(\exists c)(\forall I \in \mathcal{I})(\forall S \in I)$

$$K_{\Psi}(S \mid I) + m_T(S) < c + \sup_{\delta}[\delta + \log_2 \| T_{\delta} \cap I \|].$$

(b) $(\forall T \in \mathcal{T})(\exists c)(\forall I \in \mathcal{I})$

$$\| I \| < \infty \Rightarrow (\exists S \in I)\, K_{\Psi}(S \mid I) < c,\, m_T(S) = \min_{S' \in I} m_T(S').$$

The application of Theorem 3(a) will be to show that, for properly selected $\mathcal{I}$, sequences of high conditional complexity correspond to those of low m_T, and would therefore tend to pass a statistical test T. The application of Theorem 3(b) will be to show that there is in general no converse to the preceding; there may exist sequences passing a statistical test that are of low complexity. We will find that the complexity-based definitions are more stringent than the corresponding statistical definitions or tests of a probabilistic model or hypothesis.

VE. Definition of Random Finite Sequence Using Complexity

The usual probabilistic description of a random sequence is that it is the outcome of a random experiment. The ensemble of experimental outcomes (finite-length binary sequences) is assigned a probability distribution such that all sequences of the same length are equally probable,

$$\Theta = \{1, 2, \dots\}, \qquad P_{\theta}(S) = \begin{cases} 2^{-\theta} & \text{if} \quad l(S) = \theta \\ 0 & \text{if} \quad l(S) \neq \theta. \end{cases}$$

There is no distinction between regular and irregular outcomes. Nevertheless, in practice, a regular outcome of a supposedly random experiment is considered suspect. We would tend to reject the hypothesis of, say, random tosses of a fair coin if we observed the outcome to be a long (e.g., 10 trials) run of heads. Taking practice as our guide, we would expect that a sequence should be highly irregular or complex for it to be considered as random. In view of the discussion in Subsection VD3 it seems appropriate to take $\mathcal{I} = \{\{S : l(S) = j\}\}$ and introduce

Definition 7. The sequence S is random if for preselected (Ψ, t), with Ψ an AUTM,

$$K_{\Psi}(S \mid l(S)) > l(S) - t.$$

A partial justification for this definition is available from the verifications of A2 and A3 in Subsection VC3. It was shown there that:

(1) $(\exists c)(\forall S)\, K_{\Psi}(S \mid l(S)) < l(S) + c;$

(2)   there are at most 2^{l-t+1} sequences of length l which are not random (Ψ, t).

Hence maximally complex sequences (the "most random" sequences) have complexity within a constant of their length. The larger t is, the less random is the sequence. For large t there are only a few sequences; they are easily described; and they are not random.

We can also justify Definition 7 by showing that sequences for which the complexity test accepts the hypothesis of randomness also tend to be sequences that are statistically judged to be random. Let $\mathcal{U} = \{U^{(i)}\}$ denote the countable collection of effective statistical tests for randomness. Definition 4(d) is satisfied under the probabilistic hypothesis of randomness if and only if

$$(\forall U \in \mathcal{U})(\forall \delta)(\forall n) \| U_{\delta} \cap \{S : l(S) = n\}\| \leqslant 2^{n-\delta}.$$

With this understanding as to what constitutes a statistical test for the probabilistic hypothesis, we can assert

Theorem 4(a).

$$(\forall U \in \mathcal{U})(\exists c)(\forall S)\, K_{\Psi}\big(S \mid l(S)\big) + m_U(S) < l(S) + c.$$

Hence,

$$K_{\Psi}\big(S \mid l(S)\big) > l(S) - t \Rightarrow m_U(S) < t + c,$$

and by Definitions 5 and 7 for statistical and complexity acceptance regions we see that long sequences judged random by the complexity approach will also tend to be judged random by the statistical approach. The absence of a converse is established through

Theorem 4(b).

$$(\forall U \in \mathcal{U})(\exists c)(\forall n)(\exists S)\, l(S) = n,\, K_{\Psi}\big(S \mid l(S)\big) < c,$$

$$m_U(S) = \min\{m : l(S') = n,\, m_U(S') = m\}.$$

Thus the complexity-based definition is more stringent than the statistical definition of randomness. We judge this to the credit of the complexity approach.

Is conditional complexity as thus far defined a sufficient basis for the definition of a random sequence? It might be argued that complexity suffices only for an appropriate ("reasonable") AUTM, although we are unable to present such an argument. Furthermore, we know of no guidelines for the selection of the degree of randomness at which to draw the line between random and nonrandom sequences. As we have seen, there is a corresponding indeterminacy in the statistical definition of a random sequence, involving the choice of critical level. To arrive at a complexity definition we must make what at present appear to be arbitrary choices.

VF. Random Infinite Sequences

1. Complexity-Based Definitions

Random infinite sequences are of no practical concern, although they are an interesting idealization. A straightforward extension of the complexity definition of a random finite sequence would be as follows.

The infinite sequence S is random if, when S_n denotes an initial segment of S of length n,

$$(\exists c)(\forall n)\, K_\Psi(S_n \mid n) > n - c.$$

Unfortunately, this definition is vacuous; there do not exist such infinite sequences. Martin-Löf [14] has shown

Theorem 5. If $\{g_n\}$ such that there exists a computable sequence $\{f_n\}$ with $f_n \geqslant g_n$ and $\sum_{n=1}^{\infty} 2^{-f_n} = \infty$, then there does not exist an infinite sequence S such that

$$(\forall n)\, \big(K_\Psi(S_n \mid n) > n - g_n\big).$$

A corollary is $(\forall S)(\exists c)(\forall n)\, K_\Psi(S_n \mid n) < n - \log \log n + c$. A converse to Theorem 5 is

Theorem 6. If $\{f_n\}$ such that $\sum_{n=1}^{\infty} 2^{-f_n} < \infty$, then for almost all (Bernoulli measure, $p = \frac{1}{2}$) infinite sequences

$$(\exists N)(\forall n > N)\, \big(K_\Psi(S_n \mid n) > n - f_n\big).$$

A variety of definitions for an infinite random sequence S have been proposed, and they are not equivalent. Three of these definitions are

Kolmogorov [15]. $(\exists c)(\forall n)(\exists k \geqslant n, S_k') K_\Psi(S_k' \mid k) > k - c$, S_n is an initial segment of S_k'.

Loveland–Martin-Löf [16, p. 44]. $(\exists c)(K_\Psi(S_n \mid n) > n - c$, infinitely often.

Chaitin [8, p. 147].   Select $f_n \geqslant 3 \log_2 n$. Then

$$(\exists n)(\forall m > n)(K_\Psi(S_m \mid m) > m - f_m).$$

Results on the interrelations between these definitions (e.g., Loveland–Martin-Löf implies Kolmogorov) and those of the collective are available in the work of Bruere-Dawson [16].

A somewhat different approach to the definition of an infinite random sequence that is still in the spirit of recursive function theory, though not of computational complexity, has been described by Martin-Löf [17]. The set of infinite random sequences is defined to be the intersection of all hyperarithmetical sets of measure 1. Equivalently, a random sequence has all of the properties expressible in the constructive infinitary propositional calculus that are with probability 1 properties of probabilistic random sequences (e.g. $\lim_{n \to \infty} w(S_n)/n = \frac{1}{2}$). The reader interested in this development, of value only for infinite sequences, should consult [17].

2. Statistical Definition

While there are difficulties in generating a complexity-based definition of a random infinite sequence, it is possible to extend the idea of a statistical test to generate a statistical definition of a random infinite sequence. For a statistical definition we should like to assign a critical level $m_U(S)$ to an infinite sequence S. Any practical test for randomness must be such that the decision is reached by examination of only a finite initial segment S_n of S. Once an initial segment is rejected at a significance level $2^{-\delta}$, then all longer initial segments, and hence the infinite sequence S, must also be rejected as random at that significance level. Tests with this property are called sequential tests, and they satisfy

Definition 8. U is a sequential test for randomness if $U \in \mathscr{U}$ and $(\forall n \geqslant m) S_m \in U_\delta \Rightarrow S_n \in U_\delta$, where S_k denotes an initial segment of length k of an infinite sequence S.

Note that by the construction of a sequential test U, the critical level $2^{-m_U(S_n)}$ of initial segments must be nonincreasing in n. From this monotonicity property and the definition of the critical level $2^{-m_U(S)}$ for the infinite sequence S as the smallest significance level at which S is rejected we see that

$$m_U(S) = \lim_{n \to \infty} m_U(S_n).$$

The extension of the statistical definition of random finite sequences suggested by Martin-Löf is given by

Definition 9.   The infinite sequence S is U-random if $m_U(S)$ is finite.

We need no longer specify an arbitrary threshold t to define a random infinite sequence.

3. Relations between the Complexity and Statistical Definitions

The basic result is

Theorem 7.   (a)   For all sequential tests U, if S is random in the sense of Kolmogorov or of Loveland and Martin-Löf, then S is U-random.
   (b)   It is false in general that if S is U-random, then it is also random in either the Kolmogorov or Loveland and Martin-Löf sense.

A disadvantage of the Chaitin definition is that it does not imply U-randomness.
   For recent work on the concept of an infinite random sequence based upon Ville's martingales see Schnorr [18].

VG. Exchangeable and Bernoulli Finite Sequences

1. Exchangeable Sequences

The probabilistic hypothesis of exchangeable (or symmetric) finite sequences is described by

$$(\forall P \in \{P_\theta , \, \theta \in \Theta\})(\forall S = x_1 \cdots x_{l(S)})(\forall \{i_j\} \text{ permutation of } \{j\})$$

$$(S' = x_{i_1} \cdots x_{i_{l(S)}}) \, P(S) = P(S').$$

Somewhat more constructively, we have from the de Finetti characterization of exchangeable measures (Section IVB) that

$$(\forall P \text{ exchangeable}) \left(\exists \{\lambda_i\}, \lambda_i \geqslant 0, \sum \lambda_i \leqslant 1 \right) \left(\exists \{(n_i , w_i)\} \right)$$

$$(\exists F \text{ distribution function with support } [0, 1])$$

$$P(S) = \left(1 - \sum \lambda_i \right) \int_0^1 p^{w(S)} (1 - p)^{l(S) - w(S)} \, dF(p)$$

$$+ \begin{cases} \lambda_i \Big/ \binom{n_i}{w_i} & \text{if} \quad (\exists i) \, l(S) = n_i , w(S) = w_i \\ 0 & \text{otherwise,} \end{cases}$$

where $w(S)$ is the number of 1's in S. Note that all exchangeable sequences sharing $(l(S), w(S))$ have the same probability. In view of Subsection VD3 this suggests taking

$$\mathscr{I} = \{\{S : l(S) = n, w(S) = w\}\}$$

as a basis for a maximal conditional complexity definition of, or test for, exchangeability.

Definition 10. S is complexity-sense exchangeable if, for preassigned AUTM Ψ and threshold t,

$$K_\Psi(S \mid l(S), w(S)) > \log_2 \binom{l}{w} - t.$$

As was the case in the definition of random sequences, it is practice and not probability theory that suggests associating maximal conditional complexity with exchangeability. To further illuminate the complexity-based definition or test we compare it with statistical definitions of exchangeability.

If we let θ index a choice of F, λ, and sequence length l, then the family $\mathscr{E}$ of statistical tests for exchangeability satisfying Definition 4(d) is such that

$$(\forall E \in \mathscr{E})(\forall \delta) \, \| E_\delta \cap \{S : l(S) = n, w(S) = w\} \| \leqslant 2^{-\delta} \binom{n}{w};$$

this is easily verified from the de Finetti characterization of exchangeable measures given above and the significance level constraint

$$\sup_{\theta \in \Theta} P_\theta(E_\delta) \leqslant 2^{-\delta}.$$

The relation between the complexity $K_\Psi(S \mid l, w)$ and critical level $2^{-m_E(S)}$ of exchangeable sequences is given by

Theorem 8.

(a) $(\forall E \in \mathscr{E})(\exists c)(\forall S) \, K_\Psi(S \mid l(S), w(S)) + m_E(S) < \log_2 \binom{l}{w} + c;$

(b) $(\forall E \in \mathscr{E})(\exists c)(\forall n, w)(\exists S) \, l(S) = n, w(S) = w, K_\Psi(S \mid l, w) < c,$
$m_E(S) = \min\{m : l(S') = n, w(S') = w, m_E(S') = m\}.$

It is immediate from Theorem 8(a) that sequences judged exchangeable by the complexity-based definition also tend to be judged as statistically exchangeable,

$$(\forall E \in \mathscr{E})(\exists c)(\forall t)(\forall S) \, K_\Psi(S \mid l, w) > \log_2 \binom{l}{w} - t \Rightarrow m_E(S) < t + c.$$

Theorem 8(b) assures us that the converse implication is in general false. The complexity-based concept of exchangeability is more stringent than the statistical one.

2. Bernoulli Sequences

It is informative concerning the limitations of this complexity approach to consider the subclass of exchangeable distributions that correspond to the Bernoulli model,

$$P_\theta(S) = p^{w(S)}(1 - p)^{l(S)-w(S)}.$$

Following the discussion in Subsection VD3 we would presumably adopt exactly the same complexity-based definition for a Bernoulli sequence as was given in Definition 10 for an exchangeable sequence. The family $\mathscr{B}$ of statistical tests for the Bernoulli hypothesis satisfy Definition 4(d) if and only if

$$(\forall B \in \mathscr{B})(\forall \delta) \, \| \, B_\delta \cap \{S : l(S) = n, w(S) = w\} \|$$
$$\leqslant \min\left\{ \binom{n}{w}, 2^{-\delta} \frac{n^n}{w^w(n - w)^{n-w}} \right\}.$$

From Theorem 3(a) and elementary calculations we find that

$$K_\Psi(S \mid n, w) + m_B(S) < c + \log_2\left[\frac{n^n}{w^w(n - w)^{n-w}} \right].$$

However, since by Lemma 2

$$K_\Psi(S \mid n, w) < c' + \log_2 \binom{n}{w},$$

we see that this application of Theorem 3(a) is in general uninformative. In fact we find

Theorem 9.

$$(\exists B \in \mathscr{B})(\forall c, t)(\exists n)(\forall l > n)(\exists S)\, K_\Psi(S \mid l, w) > \log_2 \binom{l}{w} - t,\, m_B(S) > c.$$

Thus seemingly suitably conditioned maximal complexity no longer ensures satisfaction of statistical tests for Bernoulli sequences.

VH. Independence and Complexity

1. Introduction

As we remarked in the preceding chapters, the concept of independence of events or of experiments is a fundamental relation of great importance in applications of probability theory. In selecting probability models for random experiments we often rely upon prior judgments that certain events are independent to restrict the class of acceptable models; for example, the belief that successive tosses of a coin are independent reduces the family of probability models to the one-parameter Bernoulli model. In applications to such areas as forecasting, communications, and quality control we often accept the judgment, based on a selected probability model, that we cannot, say, do better than a particular prediction because the prediction error is independent of our data on the past (a Bode–Shannon or orthogonal projection type of argument). Hence we use independence both to assist us in modeling random phenomena and to indicate limitations on what we can achieve in controlling such phenomena.

Stochastic independence (SI), the product factorization of joint probabilities

$$A(\text{SI})\, B \Leftrightarrow P(A \cap B) = P(A)\, P(B),$$

is the usual formalization of the informal concept of independence, unlinkedness, or unrelatedness. The first difficulty with SI is that it presupposes an advanced state of knowledge concerning the occurrence of the events in question. Given our views as to the problems encountered in assessing probability, we do not favor a purely probability-based definition. Furthermore, such a concept of independence would be of no use to us in selecting probability models. The second difficulty with SI concerns its adequacy as a formalization. The computational-complexity-based approach that we will develop, when combined with a relative-frequency interpretation of probability, suggests that the usual definition of SI is necessary but not sufficient for a characterization of our informal or intuitive idea of independence. There can be a link between events despite the appearance of product factorization of relative-frequencies. It may well be that independence of events can no more be adequately formulated in probability terms than can disjointness of events.

2. Relative-Frequency, Complexity, and Stochastic Independence

To a relative-frequentist the meaning of $A(\text{SI})B$ is supplied, as discussed in Section IIIG, through the following construction. A compound experiment is generated through unlinked repetitions of the experiment that generates the events A and B. Binary occurrence records, O_A and O_B, are generated from the results of the performance of the compound experiment through

$$O_A = x_1 \cdots x_n,$$

$$O_B = y_1 \cdots y_n,$$

$$x_i = \begin{cases} 1 & \text{if } A \text{ occurred on trial } i \\ 0 & \text{if } A \text{ did not occur on trial } i, \end{cases}$$

$$y_i = \begin{cases} 1 & \text{if } B \text{ occurred on trial } i \\ 0 & \text{if } B \text{ did not occur on trial } i. \end{cases}$$

The statement $A(\text{SI})B$ is then held to mean that

$$\frac{1}{n} \sum_{i=1}^{n} x_i y_i \approx \left(\frac{1}{n} \sum_{i=1}^{n} x_i \right) \left(\frac{1}{n} \sum_{i=1}^{n} y_i \right),$$

corresponding to

$$P(A \cap B) = P(A) P(B).$$

If, as is common, it is deemed necessary to justify the hypothesis of unlinked repetitions, then the relative-frequentist would examine occurrence records for k-tuples of occurrences and nonoccurrences of A and of B. For example, we might be interested in whether the pair of outcomes in which occurrence of A is immediately followed by its nonoccurrence is independent of the pair of outcomes in which the occurrence of B is again followed by its occurrence. This interest would lead to new occurrence records

$$O_A^{10} = u_1 \cdots u_{n/2}, \qquad O_B^{11} = v_1 \cdots v_{n/2},$$

$$u_i = \begin{cases} 1 & \text{if} \quad x_{2i} = 1, \quad x_{2i+1} = 0 \\ 0 & \text{otherwise,} \end{cases}$$

$$v_i = \begin{cases} 1 & \text{if} \quad y_{2i} = y_{2i+1} = 1 \\ 0 & \text{otherwise.} \end{cases}$$

The pairs of outcomes would be judged SI if

$$\frac{2}{n} \sum_{i=1}^{n/2} u_i v_i \approx \left(\frac{2}{n} \sum_{i=1}^{n/2} u_i \right) \left(\frac{2}{n} \sum_{i=1}^{n/2} v_i \right).$$

If the repeated experiments were indeed unlinked, then we should find approximate product factorization of the relative-frequency for most k-tuples, so long as k is small compared to n.

Empirically the events A and B would be judged independent if O_A and O_B were thought to be unlinked sequences representing the outcomes of unlinked repetitions of a random experiment. Given only O_A and O_B, we wish to test the hypothesis that events A and B are independent. Of course, the probabilistic model for (O_A, O_B) depends as well on the manner in which the experiment generating these events is repeated. Rather than require independent and identically distributed trials we will more generally model (O_A, O_B) as a pair of independent, exchangeable sequences of equal length. Specifically,

$$(\forall P \in \{P_\theta, \theta \in \Theta\}) (\exists P_1, P_2 \text{ exchangeable distributions})$$

$$(\forall S_1, S_2) P(S_1, S_2) = P_1(S_1) P_2(S_2).$$

Following the discussion of Subsection VD3 we base a complexity test for independence on the collection

$$\mathscr{I} = \{\{S_1, S_2\} : l(S_1) = l(S_2) = n, w(S_1) = w_1, w(S_2) = w_2\}.$$

Definition 11. (S_1, S_2) is a pair of independent exchangeable sequences in the complexity sense if, for preselected AUTM Ψ and threshold t,

$$K_\Psi(S_1 S_2 \mid n, w_1, w_2) > \log_2 \left[\binom{n}{w_1} \binom{n}{w_2} \right] - t,$$

where $S_1 S_2$ is the concatenation of the sequences S_1 and S_2.

Once again we can defend Definition 11 through a comparison with statistical tests. The family $\mathscr{E}'$ of tests for the hypothesis of pairs of independent exchangeable sequences is such that

$$(\forall E' \in \mathscr{E}') \| E_\delta' \cap \{(S_1, S_2) : l(S_1) = l(S_2) = n, w(S_1) = w_1, w(S_2) = w_2\}\|$$

$$\leqslant 2^{-\delta} \binom{n}{w_1}\binom{n}{w_2}.$$

A pair of sequences (S_1, S_2) passes a test E' if, for some preassigned threshold t',

$$m_{E'}(S_1, S_2) < t'.$$

A relation between complexity-based and statistical determinations of independent, exchangeable sequences is provided by

Theorem 10.

(a) $(\forall E' \in \mathscr{E}')(\exists c) \left(\forall(S_1, S_2)\, l(S_1) = l(S_2) = n, w(S_1) = w_1, w(S_2) = w_2 \right)$

$$K_\Psi(S_1 S_2 \mid n, w_1, w_2) + m_{E'}(S_1, S_2) < \log_2 \left[\binom{n}{w_1}\binom{n}{w_2} \right] + c;$$

(b) $(\forall E' \in \mathscr{E}')(\exists c)(\forall n, w_1, w_2)\big(\exists(S_1, S_2)\, l(S_1) = l(S_2) = n,$

$$w(S_1) = w_1, w(S_2) = w_2\big)\, K_\Psi(S_1 S_2 \mid n, w_2) < c,$$

$$m_{E'}(S_1, S_2) = \min\{m : l(S_1') = l(S_2') = n,$$

$$w(S_1') = w_1, w(S_2') = w_2, m_{E'}(S_1', S_2') = m\}.$$

Thus,

$$(\exists c)\, K_\Psi(S_1 S_2 \mid n, w_1, w_2) > \log_2 \left[\binom{n}{w_1}\binom{n}{w_2} \right] - t \Rightarrow m_{E'}(S_1, S_2) < c + t,$$

and complexity-based independence implies statistical independence. Theorem 10(b) indicates the failure of the converse and the greater stringency of the complexity test.

3. Empirical Independence

In view of the preceding discussion, we are led to consider a complexity-based concept of empirically independent (EI) events $A(\text{EI})B$. In order to preserve the symmetry of the notion of independence $[A(\text{EI})B \Leftrightarrow B(\text{EI})A]$ we propose

Definition 12. $A(\text{EI})B$ if the occurrence records (O_A, O_B) satisfy, for preselected AUTM Ψ and threshold t,

$$\min[K_\Psi(O_A O_B \mid l(O_A) = l(O_B) = n, w(O_A) = w_A, w(O_B) = w_B),$$

$$K_\Psi(O_B O_A \mid n, w_A, w_B)] > \log_2\left[\binom{n}{w_A}\binom{n}{w_B}\right] - t.$$

While Theorem 10 assures us that judgments of $A(\text{EI})B$ tend to agree with statistical judgments of $A(\text{SI})B$, additional insight into the formal similarities and discrepancies between EI and SI is provided by

Theorem 11. (a) $A(\text{EI})B \Leftrightarrow B(\text{EI})A$.
  (b) $(\exists c)(\forall t > c)(\forall n)(\forall A)$ at least the fraction $1 - 2^{-(t-c)}$ of all possible occurrence records of length n are such that $A(\text{EI})\varnothing$.
  (c) $(\exists c)(\forall t > c)(\forall n)$ at least the fraction $1 - 2^{-(t-c)}$ of all possible occurrence records of length n for which $A(\text{EI})B$ also imply $A(\text{EI})\bar{B}$.
  (d) $(\forall t)(\exists \theta)(\forall n) A(\text{EI})B, A \supseteq B \Rightarrow w(O_A) > n - \theta\sqrt{n}$ or $w(O_B) < \theta\sqrt{n}$.

Theorem 11(a) assures us that EI is a symmetric relation. Theorem 11(b) suggests that in most cases, any event is independent of the null event. Theorem 11(c) suggests that EI has the complementation property of SI (axiom I3 of Section IIF), at least in most cases. If we desire I3 to hold uniformly, then we can modify Definition 12 in an obvious manner to include the conditional complexity of the concatenated sequences $O_A O_{\bar{B}}, O_{\bar{A}} O_B, O_{\bar{A}} O_{\bar{B}}, O_B O_{\bar{A}}, O_{\bar{B}} O_A, O_{\bar{B}} O_{\bar{A}}$. Parts (b) and (c) also suggest constants that the threshold t should considerably exceed. Part (d) informs us that $A(\text{EI})B$ and $A \supseteq B$ only if either A is like Ω in occurring on almost all trials or B is like $\varnothing$ in almost never occurring.

Perhaps the most significant practical discrepancy between SI as statistically determined and EI is evident from Theorem 10(b). There exist pairs of arbitrarily long occurrence records (O_A, O_B) for which a relative-frequentist would tend to believe that the events A and B are stochastically independent and conclude with $A(\text{SI})B$, yet the occurrence of A may be easily predictable from that of B.

4. Conclusions

The above-mentioned discrepancy between EI and SI occurs no matter which statistical test we employ for SI, and it suggests that independence cannot be adequately embedded in probability. This conclusion is somewhat surprising when viewed from the conventional perspective, albeit there is a parallel with set disjointness; if A and B

are disjoint, then $P(A \cup B) = P(A) + P(B)$, although from this relation between probabilities we could not conclude that A and B need be disjoint. The complexity-based approach to independence has the advantage of closely following our intuition and requiring a minimum of auxiliary hypotheses, and we are inclined to favor either this approach or the pragmatic approach mentioned in Section IIH, as explicating our informal concept of independence.

While we have only treated pairwise independence, our discussion is easily extended to mutually independent events and thence to independent experiments [19].

VI. Complexity-Based Approaches to Prediction and Probability

1. Introduction

Attempts to derive a principle to enable us to extrapolate from, say, the initial segment of a sequence to its continuation will either require some hypothesis about the source of the sequence or a very specific definition of what we mean by the activity of extrapolation. Two popular and useful metaphysical principles for the definition of extrapolation are those of simplicity (Ockham's razor) and indifference. The principle of simplicity asserts that the "simplest" explanation is the most reliable. The principle of indifference suggests that in the absence of grounds enabling us to choose between explanations, we should treat them as equally reliable.

The notion of an explanation of data, modeled as a binary sequence D, can be formalized as a program P which together with $l(D)$ computes D on a suitable universal computer, say an AUTM ϕ. We also require that if we view the data as a time series then the explanation P when combined with n should generate the initial segment D_n of length n of D,

$$\phi(P, n) = D_n .$$

The simplicity or complexity of an explanation is perhaps adequately reflected by the complexity $K_\Psi(P)$ of a program for D.

Given a data sequence D we may be interested in determining any or all of the following:

(1) best extrapolation H^* of D;

(2) relative likelihoods of two extrapolations H_1 and H_2 ;

(3) probability of a particular extrapolation H.

From our discussion in Section VC concerning the adequacy criterion A1 and Loveland's uniform complexity $K_\Psi(S; l(S))$, we see that, to within a universal constant, the simplest program P^* generating DH^* given $l(DH^*)$ satisfies

$$l(P^*) \approx K_\Psi(D; l(D)).$$

However, this formulation of the optimum extrapolation or prediction problem leaves much to be desired; the procedure is not effectively computable, there is no account taken of the consequences of errors or correct predictions, and, in general, there will not be a unique prediction. We can either examine this prediction problem in greater detail and supplement the principles of indifference and simplicity so as to remove these deficiencies, or postpone the selection of a best prediction in favor of first assessing the probabilities of all possible predictions. While we favor the former approach, we nevertheless feel it important to examine the latter one.

2. Comparative Probability

A CP order relation between sets of sequences can be represented through a suitable program on an appropriate universal machine Ψ as follows. Let the sample space Ω consist of all infinite-length binary sequences and let the collection $\mathscr{F}$ be a field containing all subsets of Ω that are cylinder sets with finitely many coordinates as base. The class of effective CP relations $\succsim$ is comprised of those for which there exists a program $P_\succsim$ such that for all $A, B \in \mathscr{F}$ described by programs P_A and P_B,

$$\Psi(P_\succsim, P_A, P_B) = \begin{cases} 1 & \text{if } A \succsim B \\ 0 & \text{if } A \prec B. \end{cases}$$

The requirement that $\succsim$ be a CP relation satisfying at least C0–C4 imposes restrictions on possible programs $P_\succsim$, although these restrictions are not easily interpreted in computational terms.

The principles of indifference and simplicity are by themselves insufficient to delineate the "good" CP relations to be used in extrapolating sequences or in making decisions based on observations. One way to supplement these principles is through principles of invariance. Illustrations of invariance arguments as applied to classical and to logical probability can be found in Sections VIC and VIID. We will present two invariance axioms for complexity-based comparative probability which seem reasonable but whose consequences have not been fully explored. The first axiom corresponds to the probabilistic hypothesis of stationarity.

1. $\gtrsim$ is shift invariant; that is, if A and B are cylinder sets with bases $\{a_{i_j}\}$ and $\{b_{i_j}\}$, and $A^{(n)}$ and $B^{(n)}$ are cylinder sets with bases $\{a_{i_j+n}\}$ and $\{b_{i_j+n}\}$, then $A \gtrsim B \Leftrightarrow A^{(n)} \gtrsim B^{(n)}$.

Axiom 1 is justifiable if we assume that the process we are observing has been operating infinitely long before we started to observe it at some time origin of our choosing that is unrelated, insofar as we know, to the process. The second axiom suggests that "similar" observations should eventually yield cylinder sets that are CP equivalent. A rather strong notion of similar infinite sequences is that, given a suitable machine ϕ, two infinite sequences A and B, with initial segments of length n denoted by A_n and B_n, are similar if there exist finite-length programs P and Q such that [6, p. 124]

$$(\forall n)\,\phi(P, A_n, n) = B_n, \qquad \phi(Q, B_n, n) = A_n.$$

We now propose

2. If the infinite sequences A and B are similar, then

$$(\exists N)(\forall n > N)\{S : S_n = A_n\} \approx \{S : S_n = B_n\}.$$

A weaker version of (2) is

2'. $(\forall A_n, B_n, S)(\exists N)(\forall m > N)\{S' : S'_{m+n} = A_n S_m\} \approx \{S' : S'_{m+n} = B_n S_m\}.$

A partial justification for this second axiom is that, insofar as our prior knowledge concerning the generation of the sequence we are observing is minimal, we should not draw distinctions between essentially equivalent observations. Some understanding of the implications of 2' is available from

Theorem 12. (a) If C0–C4 and 2', then there is a unique almost agreeing probability measure P given by

$$(\forall A_n)\, P(A_n) = 2^{-n}.$$

(b) If in addition C7, then the almost agreeing P agrees with $\gtrsim$.

It is of interest to note that axiomatic derivations leading to a uniform prior distribution have been criticized as not enabling us to learn from experience. No matter which finite sequence we observe, all continuations are equally probable. This has occasionally prompted the rejection of such axiomatizations. However, by considering CP rather then restricting ourselves to Kolmogorov-type quantitative probability, we may be able to avoid both horns of the dilemma.

3. Solomonoff's Definitions of Probability

The room for arbitrary decisions is greatly increased when we attempt to develop quantitative probability. The question of quantitative probability based on complexity was first raised and treated by Solomonoff [20]. Subsequently, Willis [21] made contributions to it, and Kolmogorov has indicated he expects to have results on this question. One of Solomonoff's definitions of quantitative probability relies heavily upon simplicity and indifference. We calculate $P(H \mid D)$ by considering all continuations C_m of length m beyond H and C_m' beyond D (to achieve a normalization), weighting continuations with short programs more heavily than those with long programs, and weighting equal-length programs equally. Specifically if H has length n, and we use the notation $(P)_i$ for the ith program for P given D, then

$$P(H \mid D) = \lim_{\epsilon \to 0} \lim_{m \to \infty} \frac{\sum_{C_m} \sum_{i=1}^{\infty} ((1 - \epsilon)/2)^{l((HC_m)_i)}}{\sum_{C_{n+m}} \sum_{i=1}^{\infty} ((1 - \epsilon)/2)^{l((C_{n+m})_i)}}.$$

The factor of $1 - \epsilon$ is introduced to ensure convergence. The denominator is intended to normalize $P(H \mid D)$ so that it has the properties of a probability distribution. In all, this is an arbitrary and contrived definition of a numerical probability; if, indeed, it defines the usual kind of quantitative probability at all.

Solomonoff has suggested three other definitions of quantitative probability and has argued that they are roughly equivalent [22]. His fourth definition is of interest in that it proceeds from a novel viewpoint. Rather than assign probability to an extrapolation through some weighting of possible models (programs) for the generation of the data, we consider "all" of the possible methods for evaluating probability. The probability we derive is formed as a weighted combination of probabilities assessed by different methods.

A probability evaluation method (PEM) is an algorithm for the assignment of a computable number $p(S)$ to each finite binary (for the purposes of simplicity) sequence S. It has the following three properties:

$$p(S) \geqslant 0, \qquad p(0) + p(1) = 1, \qquad p(S0) + p(S1) = p(S).$$

If we fix on a specific UTM Ψ, then each PEM is represented by some program P which together with the sequence S forms the input to Ψ,

$$p(S) = \Psi(P, S).$$

There are countably many distinct PEM $\{p_j(S)\}$. Each p_j can be described by countably many programs on Ψ. In order to arrive at a probability assignment p^* we take the average of the $p_j(S)$ each weighted by a factor w_j that can depend on the density and/or length of programs for p_j

$$p^*(S) = \sum_j w_j p_j(S).$$

Once choice for $\{w_j\}$, suggested by Solomonoff, is

$$w_j = \frac{\lim_{n \to \infty} (\text{number of programs for } p_j \text{ of length} < n)/2^n}{\sum_k \lim_{n \to \infty} (\text{number of programs for } p_k \text{ of length} < n)/2^n}.$$

Notice that there is no contradiction between first enumerating all PEM and then constructing a new method of probability assignment. Whereas the p_j are effective assignments, p^* is incomputable; that is, p^* cannot be expressed by a finite program on a UTM. The loss of computability occurs in the evaluation of the weights $\{w_j\}$.

4. Critique of the Complexity Approach to Probability

Difficulties with the complexity-based approaches to quantitative probability include definitions that are not effectively computable; dependence on an arbitrary choice of computer Ψ; arbitrariness in the proposed definitions; and nature of a justification for the use of such probabilities.

It has been argued that incomputability is the hallmark of any general inductive procedure [22, 23]. The essence of the argument seems to be that if there existed a generally applicable prediction (induction) procedure that was computable, then we could design a data source containing the prediction algorithm and operating, after inspection of the "internal" prediction based on its previous outputs, in such a manner as to frustrate the prediction. The argument is not fully convincing. When the inductive procedure evaluates probabilities rather than future outcomes, then the possibility of frustrating the prediction depends on the meaning of the calculated probability.

The problem of the incomputability of complexity-based probability suggests the need for approximation. In effect the series of approximations become the working definition and replace the idealized, incomputable proposals. Willis [21] has treated this question at some length. The approximations he suggests are generated by applying computation bounds. For example, we can effectively approximate complexity and the weights $\{w_j\}$ if we reject as a program any sequence for which computation takes in excess of T operations; in this way we circumvent the halting problem. Another bound is obtained by limiting the tape working space so that a calculation either halts with all calculations having occurred within the allotted space, or halts in an attempted transfer beyond the allotted space, or loops by repeating a tape and internal state configuration.

As pointed out earlier, actual complexity evaluations depend on the choice of basis machine. While different AUTM may lead to "similar" evaluations for certain problems, "similar" does not mean identical. It is easy to invent, and difficult to exclude cogently, examples for which probability assessments vary radically with the choice of machine. What is a "reasonable variant"? Why restrict ourselves only to UTM? Might there not exist physical (e.g., quantum mechanical) systems that cannot be precisely modeled by a UTM. If so, then unless we extend our class of computers we may be unable, in principle, to infer accurately a probabilistic model for such physical systems.

Solomonoff has proposed several definitions for quantitative probability and many more are possible, for example, through different choices of $\{w_j\}$ in p^*. He believes these definitions to be essentially equivalent, but there is much leeway in "essentially" and not enough in the way of rigorous, detailed supporting analysis. For example, is the notion of a universal PEM p^* in conflict with notions of independent sequences and the factorization of probability? It is conceivable that for the most heavily weighted p_j

$$p_j(SS') = p_j(S)p_j(S'),$$

yet false that

$$p^*(SS') = p^*(S)p^*(S').$$

Much more investigation is needed before we can claim a computational-complexity-based definition of probability. Even the definition of the simpler concept of comparative probability admits of more than one seemingly reasonable definition.

Finally, of course, there is the vital question of the inductive significance of complexity-based probability. It may prove possible to vindicate, though not to validate, some complexity-based definition of probability, that is, to defend its use rather than to guarantee its propensity for truth. Perhaps it can be argued that if we mean anything by induction or extrapolation, then some complexity definition is appropriate. For instance, we do not claim that $P(H \mid D)$ is the relative-frequency in some hypothetical ensemble with which D is followed by H. Rather we are encouraged to believe that there is a "rational degree of belief" in H following D that is given by $P(H \mid D)$, and this degree of belief should be the basis of forecasts. This interpretation makes the complexity approach a type of logical probability discussed in Chapter VII. However, it is unclear why complexity-based probability should be the basis of forecasting decisions. In arriving at decisions, we must arrange to take into account the consequences of these decisions, and evaluation of the consequences can involve us with the assessment of human values. How are we to combine our rational degrees of belief in outcomes with the subjective consequences of forecasts? The development of the complexity approach has, understandably, ignored this question. Furthermore the complexity-based probability assessment can only accept information or data that can be put into a finite alphabet. It is not evident that all relevant data for forecasts can be losslessly reduced to such form. The subjective approach to probability is the only approach cognizant (perhaps too much so) of this possibility. The complexity approach to probability is enlightening and promising but most of its promise is yet to be redeemed.

VJ. Reflections on Complexity and Randomness: Determinism versus Chance

The complexity characterization of random sequences seemingly has little to do with a probabilistic characterization, notwithstanding the partial agreements with statistical tests. The agreement thus far exhibited between complexity and statistics is a consequence of the assumption that a rejection region can be effectively (finitely) described and that there are arbitrarily small (low significance level) rejection regions for sufficiently long sequences. If we could identify highly complex sequences by specifying their membership in a nested sequence of sets (rejection regions), then we would have a contradiction between the asserted high complexity of the sequence and the relatively low complexity of its definition as lying in a small, easily defined set of sequences.

Of what interest, then, are the complex sequences? What is the practical relationship between random experiments and complex outcomes? We can distinguish between deterministic and chance phenomena capable of generating an indefinitely long sequence of discrete-valued outcomes on the grounds that deterministic phenomena yield outcomes of bounded complexity, whereas chance phenomena yield outcomes for which the complexity of increasingly longer outcomes diverges. Probabilistic phenomena might then be characterized as the subset of chance phenomena for which the various outcomes have apparently convergent relative-frequencies. In particular we have proven (Section IVC) that the most complex chance phenomena must be probabilistic.

What in the data or phenomena enables us to determine whether they are chance or not? From the practical viewpoint, we only observe finite outcome sequences and there is no grounds for distinction using the criteria we have suggested. It seems that to distinguish between deterministic and chance phenomena in practice would require a somewhat arbitrary definition based on apparent rates of growth of complexity or on difficult to obtain side information.

VK. Potential Applications for the Complexity Approach

The preceding discussion has revealed that the computational-complexity approach may be of value in selecting empirical definitions for such important probabilistic concepts as random, exchangeable, and independent sequences. Presumably, some other common models of probability theory are also amenable to definition from the complexity viewpoint. Indeed, to what extent can one rewrite classical probability purely from the complexity viewpoint, thereby achieving a theory of probability capable of transforming reasonable data into probability statements and classifiable as empirical and objective?

Of course, we are not just interested in theory construction and the modeling of random phenomena. What of the possibilities for new viable approaches to the basic problems of inference and decision making? Possible applications to information theory have been suggested by Kolmogorov [4], and Ziv [11] has considered the application of computational complexity to the specific area of source approximation (rate-distortion theory). Tartara [24] has provided a preliminary discussion of the use of complexity for hypothesis testing in a communications context. Solomonoff [25] has explicitly discussed the use of complexity in prediction or forecasting. These studies, however, are at best tentative, and much remains to be done to realize the potential of computational-complexity-based approaches to inference and decision-making.

V. Appendix: Proofs of Results

Lemma 1.   If Ψ is an AUTM, ϕ any TM, then

$$(\exists c)(\forall S)\left(K_\Psi(S) < c + K_\phi(S)\right),$$

where

$$K_\phi(S) = \min\{l : \phi(P) = S, l(P) = l\}.$$

Proof. ϕ being a TM is, say, the xth on the list of all TM. Hence, where defined, $\Psi(f_2(X, P)) = \phi(P)$, since Ψ is a UTM. Note that $K_\phi(S) = k$ implies that there exists P' with $l(P') = k$, $\phi(P') = S$. Therefore, $\Psi(f_2(X, P')) = S$ and $K_\Psi(S) \leqslant l(f_2(X, P'))$. However, Ψ is an AUTM, so that $K_\Psi(S) \leqslant c(X) + l(P') = c(X) + K_\phi(S)$.   ∎

Corollary.   If Ψ and ϕ are both AUTM, then

$$(\exists c)(\forall S)\left(\mid K_\Psi(S) - K_\phi(S)\mid < c\right).$$

Proof. ϕ and Ψ being TM are, say, the xth and the yth, respectively, on the list of all TM. Applying Lemma 1 yields

$$K_\Psi(S) < c_\Psi(X) + K_\phi(S), \qquad K_\phi(S) < c_\phi(Y) + K_\Psi(S),$$

where c_Ψ and c_ϕ arise from the possibly distinct encoding functions for the AUTM Ψ and ϕ. The corollary is immediate with $c = \max(c_\Psi(X), c_\phi(Y))$.   ∎

Theorem 1. K_Ψ satisfies the adequacy criteria A1–A3.

Proof.   To verify A1, note that we have to show that

$$K_\Psi(S) = \min\{k : \Psi(P) = S, l(P) = k\} \approx \theta_\Psi(S)$$

$$= \min\{k : \Psi(P) = S, k = \min\{l : \Psi(P') = P, l(P') = l\}\}.$$

Equivalently if we introduce the TM ϕ given by

$$\phi(P) = \Psi(\Psi(P)),$$

then

$$\theta_{\Psi}(S) = \min\{l : \phi(P') = S, l = l(P')\} = K_{\phi}(S).$$

Invoking Lemma 1 we see that

$$(\exists c)(\forall S)\left(K_{\Psi}(S) < c + \theta_{\Psi}(S)\right).$$

To obtain an inequality in the opposite direction assume that P^* is such that $l(P^*) = K_{\Psi}(S)$ and $\Psi(P^*) = S$. Clearly

$$\theta_{\Psi}(S) = K_{\phi}(S) \leqslant K_{\Psi}(P^*).$$

Introducing the trivial TM ω that does no computation and produces as output its input, we see that

$$K_{\omega}(S) = l(S).$$

Invoking Lemma 1 yields

$$(\exists c')(\forall S)\left(K_{\Psi}(S) < l(S) + c'\right). \tag{$*$}$$

Hence

$$\theta_{\Psi}(S) < l(P^*) + c' = K_{\Psi}(S) + c'.$$

We therefore conclude that

$$(\exists c'')(\forall S)\left(\mid K_{\Psi}(S) - \theta_{\Psi}(S)\mid < c''\right).$$

This last statement can be taken as a statement of approximate equality, and therefore, a verification of A1.

The verification of A2 is immediate from $(*)$.

Satisfaction of A3 follows from Ψ being a single-valued function. There are only 2^l distinct programs of length l. Hence there are at most 2^l sequences of complexity l. ∎

Theorem 2. $R \in \mathcal{T}$ is a universal test only if

$$(\exists\{c_i\})(\forall R_{\delta} \in R)\, R_{\delta} \supseteq \bigcup_{i=1}^{\infty} \bigcup_{\delta' \geqslant \delta + c_i} T_{\delta'}^{(i)}.$$

Proof. By Definition 6 for R universal

$$(\forall i)(\exists c_i)(\forall S)\, m_{T^{(i)}}(S) < c_i + m_R(S).$$

Noting that

$$S \in \bigcup_{i=1}^{\infty} \bigcup_{\delta' \geqslant \delta + c_i} T_{\delta'}^{(i)} \Rightarrow (\exists j, \delta'' \geqslant \delta + c_j) \, S \in T_{\delta''}^{(j)},$$

yields

$$m_{T^{(j)}}(S) \geqslant \delta'' \geqslant \delta + c_j.$$

Hence, $m_R(S) > \delta$ and $S \in R_\delta$. ∎

Theorem 3.

(a) $(\forall T \in \mathcal{T})(\exists c)(\forall I \in \mathcal{I})(\forall S \in I)$
$$K_{\Psi}(S|I) + m_T(S) \leqslant c + \sup_{\delta}[\delta + \log_2 \| T_\delta \cap I \|].$$

(b) $(\forall T \in \mathcal{T})(\exists c)(\forall I \in \mathcal{I}) \| I \| < \infty$
$$\Rightarrow (\exists S \in I) \, K_{\Psi}(S \mid I) < c, \, m_T(S) = \min_{S' \in I} m_T(S').$$

Proof. (a) For convenience denote $\sup_\delta[\delta + \log_2 \| T_\delta \cap I \|]$ by σ. Part (a) is trivially true if $\sigma = \infty$. Hence, assume $\sigma < \infty$. If we take $\delta = 0$ and note that $T_0 \cap I = I$ we see that

$$\infty > \sigma \geqslant \| I \|. \tag{$*$}$$

From the finiteness of I it is immediate that

$$(\forall S \in I)(\exists \delta_I(S)) \, I \cap T_{\delta_I(S)} = I \cap \left(\bigcap_{\delta \leqslant m_T(S)} T_\delta \right).$$

Hence, taking $\delta \uparrow m_T(S)$ to obtain a sequence of lower bounds to σ, we find

$$(\forall S \in I) \, \sigma \geqslant m_T(S) + \log_2 \| I \cap T_{\delta_I(S)} \|. \tag{$**$}$$

We construct an effective function $\phi(P, I)$ as follows. Denote the binary representation of the integer p by P. Adopt any effective enumeration of the set of finite binary sequences. Denote the integer part of χ by $\{\chi\}$. Define $\delta_I^*(P)$ by

$$I \cap T_{\delta_I^*(P)} = I \cap \left(\bigcap_{\delta < \sigma - l(P)} T_\delta \right).$$

We define $\phi(P, I)$ as the pth sequence in $I \cap T_{\delta_I^*(P)}$, if there is one, and ϕ is undefined otherwise. Note that I and T_δ are recursive sets and $\delta_I^*(P)$ is effective to conclude that ϕ is effective.

We assert that

$$(\forall S \in I)(\exists P)\, l(P) = \{\sigma\} - \{m_T(S)\} + 1, \qquad \phi(P, I) = S.$$

Given S we define P as follows. Effectively calculate $\{m_T(S)\}$ to determine that

$$S \in I \cap T_{\{m_T(S)\}}.$$

From (∗∗) we observe that

$$\| I \cap T_{\{m_T(S)\}} \| \leqslant 2^{\{\sigma\} - \{m_T(S)\} + 1}.$$

Enumerate the sequences in $I \cap T_{\{m_T(S)\}}$ to find that S is the pth sequence and represent p by P with $l(P) = \{\sigma\} - \{m_T(S)\} + 1$, as is possible. Clearly, given P, we determine $\{m_T(S)\}$ from

$$\{m_T(S)\} = \{\sigma\} + 1 - l(P)$$

and take $\phi(P, I)$ as the pth sequence in $I \cap T_{\{\sigma\} + 1 - l(P)}$.
Thus,

$$(\forall S \in I)\, K_\phi(S \mid I) \leqslant \sigma + 2 - m_T(S).$$

Since ϕ is effective, it can be simulated on Ψ at the cost of at most c' bits. Hence, invoking Lemma 1,

$$K_\Psi(S \mid I) \leqslant c + \sigma - m_T(S),$$

thereby establishing part (a).

(b)   To prove (b) we define $\phi(I)$ as the first sequence $S \in I$ for which

$$m_T(S) = \min_{S' \in I} m_T(S'),$$

and leave it undefined if there is no such sequence. Noting that m_T is effective and the minimum is taken over a finite recursive set, we see that ϕ is effective. Hence we can simulate ϕ on Ψ to conclude that $K_\Psi(S \mid I) \leqslant c$. ∎

Theorem 4(a).

$$(\forall U \in \mathcal{U})(\exists c)(\forall S)\, K_\Psi\big(S \mid l(S)\big) + m_U(S) < l(S) + c.$$

Proof.   Recall the notation of Theorem 3(a) wherein U corresponds to T and $I_n = \{S : l(S) = n\} \in \mathcal{I}$. Having observed that

$$\| U_\delta \cap I_n \| \leqslant 2^{n-\delta},$$

Theorem 4(a) becomes a corollary of Theorem 3(a). ∎

Theorem 4(b).

$$(\forall U \in \mathscr{U})(\exists c)(\forall n)(\exists S)\, l(S) = n, K_{\Psi}(S \mid l(S)) < c,$$

$$m_U(S) = \min\{m : l(S') = n, m_U(S') = m\}.$$

Proof. From $I = \{S : l(S) = n\}$ we see that $\|I\| = 2^n$. Hence Theorem 4(b) is a corollary of Theorem 3(b). ∎

Theorem 5. If $\{g_n\}$ such that there exists a computable sequence $\{f_n\}$ with $f_n \geqslant g_n$ and $\sum_{n=1}^{\infty} 2^{-f_n} = \infty$, then there does not exist an infinite sequence S such that

$$(\forall n)\,(K_{\Psi}(S_n \mid n) > n - g_n).$$

Proof. In outline we construct a computer $\phi(P, N)$ based upon a sequence $\{A_n\}$ of sets of sequences. To define $\{A_n\}$ we first agree to enumerate sequences of length n, say, in the order of the integers for which they are a binary representation. Now recursively define the infinite sequence $\{n_i\}$ by

$$n_0 = 0,$$

n_i is the smallest integer for which

$$\sum_{j=n_{i-1}+1}^{n_i} 2^{-f_j} \geqslant 1 \qquad (i > 0).$$

That this definition is possible follows from the divergence of $\sum_{j}^{\infty} 2^{-f_j}$.
    Recursively define $\{A_n\}$ by

$$A_n = \begin{cases} \text{first } 2^{n_k+1-f_{n_k+1}} \text{ sequences of length } n_k + 1, & \text{if } n = n_k + 1 \\[2mm] \text{first } 2^{n-f_n} \text{ sequences of length } n \text{ with initial segments not in any } A_j \\ \quad \text{for } n_k + 1 \leqslant j < n, & \text{if } n_{k+1} > n > n_k \\[2mm] \text{all sequences of length } n_{k+1} \text{ with initial segments not in any } A_j \\ \quad \text{for } n_k + 1 \leqslant j < n_{k+1}, & \text{if } n = n_{k+1}. \end{cases}$$

We claim that every sequence of length n_i has an initial segment of some length $n > n_{i-1}$ lying in A_n. To verify this define B_n as the set of sequences of length n_i with initial segments in A_n. Our claim is equivalent to

$$\left\| \bigcup_{n=n_{i-1}+1}^{n_i} B_n \right\| = 2^{n_i}.$$

Note that by definition of the $\{A_n\}$, the $\{B_n\}$ are disjoint and

$$\| B_n \| = 2^{n_i-n}(2^{n-f_n}) = 2^{n_i-f_n} \quad \text{for} \quad n < n_i.$$

Hence

$$\left\| \bigcup_n B_n \right\| = 2^{n_i}\left(\sum_{n=n_{i-1}+1}^{n_i-1} 2^{-f_n} + 2^{-n_i}\| A_{n_i}\| \right).$$

It is immediate from the definition of $\{A_n\}$ and $\{n_i\}$ that our claim is verified with $\| A_{n_i}\| \leqslant 2^{n_i-f_{n_i}}$. Restated we have that

$$(\forall i, S)(\exists n_{i-1} < n \leqslant n_i)\, S_n \in A_n. \tag{$*$}$$

From the recursiveness of $\{f_n\}$ follows the recursiveness of $\{n_i\}$ and of $\{A_n\}$. Define the recursive function

$$\phi(P, N) = \begin{cases} p\text{th sequence in } A_n & \text{if } p \leqslant \| A_n \| \\ (p - \| A_n \|)\text{th sequence in } \bar{A}_n & \text{if } p > \| A_n \|, \end{cases}$$

where P, N are binary representations of the integers p, n. It follows from the cardinality of A_n that

$$S_n \in A_n \Rightarrow K_\phi(S_n \mid n) \leqslant \log_2(2^{n-f_n}).$$

Hence for any AUTM Ψ we have that

$$(\exists c)(\forall n, S_n \in A_n)\, K_\Psi(S_n \mid n) < n - f_n + c. \tag{$**$}$$

Combining $(*)$ and $(**)$ yields

$$(\forall S)(\forall N)(\exists n > N)\, K_\Psi(S_n \mid n) < n - f_n + c.$$

The constant c can be absorbed in a redefinition of f_n that leaves unchanged the conditions for divergence of $\sum_{j=1}^{\infty} 2^{-f_j}$.

Finally, if $K(S_n \mid n)$ infinitely often goes below $n - f_n$, then it infinitely often goes below $n - g_n$ for $g_n \leqslant f_n$. ∎

Theorem 6. If $\{f_n\}$ such that $\sum_{n=1}^{\infty} 2^{-f_n} < \infty$, then for almost all (Bernoulli measure, $p = \frac{1}{2}$) infinite sequences

$$(\exists N)(\forall n > N)(K_\Psi(S_n \mid n) > n - f_n).$$

Proof. The proof uses a probability model to count sequences. Assign to the set of infinite binary sequences the Bernoulli $p = \frac{1}{2}$ measure. Let

$$D_n = \{S : K_\Psi(S_n \mid n) > n - f_n\}.$$

Then the theorem asserts that

$$P\left(\bigcup_{n=1}^{\infty} \bigcap_{m>n} D_m\right) = 1,$$

or equivalently,

$$P\left(\bigcap_{n=1}^{\infty} \bigcup_{m>n} \bar{D}_m\right) = 0.$$

To verify this latter conjecture, note that

$$(\forall k)\, P\left(\bigcap_{n=1}^{\infty} \bigcup_{m>n} \bar{D}_m\right) \leqslant \sum_{m=k}^{\infty} P(\bar{D}_m).$$

By the properties of K_Ψ we know that

$$\| \bar{D}_m \| < 2^{n-f_m+1}.$$

Hence

$$P(\bar{D}_m) < 2^{1-f_m},$$

and

$$(\forall k)\, P\left(\bigcap_{n=1}^{\infty} \bigcup_{m>n} \bar{D}_m\right) < 2 \sum_{m=k}^{\infty} 2^{-f_m}.$$

Invoking the hypothesis of the convergence of $\sum_1^{\infty} 2^{-f_m}$ we see that

$$P\left(\bigcap_{n=1}^{\infty} \bigcup_{m>n} \bar{D}_m\right) = 0,$$

as claimed. ∎

Theorem 7. (a) For all sequential tests U, if S is random in the sense of Kolmogorov or of Loveland and Martin-Löf, then S is U-random.

(b) It is false in general that if S is U-random, then it is also random in either the Kolmogorov or Loveland and Martin-Löf sense.

Proof. (a) Since sequences random in the sense of Loveland and Martin-Löf are also random in the sense of Kolmogorov, it suffices to verify U-randomness for the Kolmogorov sense of complexity-based randomness. If S is Kolmogorov-sense random, then

$$(\exists c)(\forall n)(\exists k \geqslant n)(\exists S_k')\, (S_n \text{ initial segment of } S_k',\, K_\Psi(S_k' \mid k) > k - c).$$

Hence by Theorem 4(a),

$$(\forall U)(\exists c')\, m_U(S_k') < c'.$$

Since U is a sequential test and S_n is an initial segment of S_k',

$$(\forall n)\, m_U(S_n) \leqslant m_U(S_k') < c',$$

and, therefore, $m_U(S) < c'$. Thus S is U-random, as asserted.

(b)   It suffices to verify (b) for the Kolmogorov sense of randomness. There exist sequences that are von Mises collectives and yet the complexity of their initial segments of length n grows as slowly as $0(\log n)$ (see [15, p. 663]). Such sequences do not satisfy the Kolmogorov definition of randomness. However, collectives have many of the generally desired relative-frequency properties of a random sequence and thus would pass some statistical tests. ∎

Theorem 8.

(a) $(\forall E \in \mathcal{E})(\exists c)(\forall S)\, K_\Psi(S \mid l(S), w(S)) + m_E(S) < \log_2 \binom{l}{w} + c;$

(b) $(\forall E \in \mathcal{E})(\exists c)(\forall n, w)(\exists S)\, l(S) = n,\, w(S) = w,\, K_\Psi(S \mid l, w) < c,$
$$m_E(S) = \min\{m : l(S') = n,\, w(S') = w,\, m_E(S') = m\}.$$

Proof.   Follows as a corollary to Theorem 3, as illustrated in the proof of Theorem 4, when we identify $\mathcal{I} = \{\{S : l(S) = n,\, w(S) = w\}\}$. ∎

Theorem 9.

$$(\exists B \in \mathcal{B})(\forall c, t)(\exists n)(\forall l > n)(\exists S)\, K_\Psi(S \mid l, w) > \log_2 \binom{l}{w} - t,\, m_B(S) > c.$$

Proof.   Define $B = \{B_\delta\}$ through

$$B_\delta = \bigcup_{n=\{2^{2\delta}+1\}}^{\infty} \{S : l(S) = n,\, w(S) = n/2\}.$$

To verify that $B \in \mathcal{B}$ it suffices to note that

$$\sup_\theta P_\theta(B_\delta) = \binom{n}{n/2} 2^{-n} \quad \text{for} \quad n = \{2^{2\delta} + 1\}.$$

From Stirlings's approximation to factorial we see that

$$\binom{n}{n/2} 2^{-n} < 1/\sqrt{n},$$

and it follows that B satisfies Definition 4(d). Hence all sequences S with $l(S) = n$, $w(S) = n/2$, for $n > \{2^{2c} + 1\}$ satisfy $m_B(S) > c$ and this includes all the maximal complexity sequences with the same length and weight. ∎

Theorem 10.

(a) $(\forall E' \in \mathscr{E}')(\exists c)(\forall(S_1, S_2)\, l(S_1) = l(S_2) = n,\, w(S_1) = w_1,\, w(S_2) = w_2)$

$$K_\Psi(S_1 S_2 \mid n, w_1, w_2) + m_{E'}(S_1, S_2) < \log_2\left[\binom{n}{w_1}\binom{n}{w_2}\right] + c;$$

(b) $(\forall E' \in \mathscr{E}')(\exists c)(\forall n, w_1, w_2)(\exists(S_1, S_2)\, l(S_1) = l(S_2) = n,$

$$w(S_1) = w_1,\, w(S_2) = w_2)\, K_\Psi(S_1 S_2 \mid n, w_1, w_2) < c,$$

$$m_{E'}(S_1, S_2) = \min\{m : l(S_1') = l(S_2') = n,\, w(S_1') = w_1,$$

$$w(S_2') = w_2,\, m_{E'}(S_1', S_2') = m\}.$$

Proof.   Both parts of this theorem are essentially corollaries to Theorem 3, and their proof can be carried out as illustrated for Theorem 4. ∎

Theorem 11.   (a) $A(\mathrm{EI})B \Leftrightarrow B(\mathrm{EI})A$.

(b) $(\exists c)(\forall t > c)(\forall n)(\forall A)$ at least the fraction $1 - 2^{-(t-c)}$ of all possible occurrence records of length n are such that $A(\mathrm{EI})\varnothing$.

(c) $(\exists c)(\forall t > c)(\forall n)$ at least the fraction $1 - 2^{-(t-c)}$ of all possible occurrence records of length n for which $A(\mathrm{EI})B$ also imply $A(\mathrm{EI})\bar{B}$.

(d) $(\forall t)(\exists \theta)(\forall n)\, A(\mathrm{EI})B,\, A \supseteq B \Rightarrow w(O_A) > n - \theta\sqrt{n}$ or $w(O_B) < \theta\sqrt{n}$.

Proof.   (a)   Immediate from the invariance of Definition 12 under interchange of O_A and O_B.

(b)   The occurrence record O_ϕ for $\varnothing$ is just the all 0's sequence. Hence,

$$(\exists c)(\forall O_A) \min\{K_\Psi(O_A O_\phi \mid n, w_A, w_\phi), K_\Psi(O_\phi O_A \mid n, w_A, w_\phi)\}$$

$$\geqslant K_\Psi(O_A \mid n, w_A) - c.$$

Referring to Definition 12, we see that it suffices for $A(\mathrm{EI})\varnothing$ that

$$K_\Psi(O_A \mid n, w_A) \geqslant \log_2\binom{n}{w_A} - (t - c).$$

Since there are fewer than a fraction $2^{-(t-c)}$ of records O_A of complexity less than $\log\binom{n}{w_A} - (t - c)$, part (b) follows.

(c)   To verify (c) note that $O_{\bar{B}}$ is derivable from O_B by termwise complementation (interchange of 0 and 1).

Hence

$$(\exists c)(\forall O_A, O_{\bar{B}}) \min\{K_\Psi(O_A O_{\bar{B}} \mid n, w_A, w_{\bar{B}}), K_\Psi(O_{\bar{B}} O_A \mid n, w_A, w_B)\}$$
$$\geq \min\{K_\Psi(O_A O_B \mid n, w_A, w_B), K_\Psi(O_B O_A \mid n, w_A, w_B)\} - c.$$

The remainder of the proof for (c) is parallel to that of (b).

(d)   To verify (d) first note that

$$A(\text{EI})B \Rightarrow K_\Psi(O_A O_B \mid n, w_A, w_B) > \log_2 \left[\binom{n}{w_A} \binom{n}{w_B} \right] - t.$$

When $A \supseteq B$, we may envision computing $O_A O_B$ by first computing O_A and then selecting the proper subset of 1's in O_A for O_B. Hence,

$$(\exists c) \, A \supseteq B \Rightarrow K_\Psi(O_A O_B \mid n, w_A, w_B) < \log_2 \left[\binom{n}{w_A} \binom{w_A}{w_B} \right] + c.$$

Thus

$$A(\text{EI})B, A \supseteq B \Rightarrow 2^{t+c} > \binom{n}{w_B} \Big/ \binom{w_A}{w_B}.$$

If for some $\theta > 0$

$$n - \theta \sqrt{n} > w_A \geq w_B > \theta \sqrt{n},$$

then

$$\frac{\binom{n}{w_B}}{\binom{w_A}{w_B}} = \prod_{k=w_A+1}^{n} \left(\frac{k}{k - w_B} \right) \geq \prod_{k=n-\theta\sqrt{n}}^{n} \left(\frac{k}{k - \theta \sqrt{n}} \right).$$

Since $k/(k - \theta\sqrt{n})$ is decreasing in k, we find that

$$\frac{\binom{n}{w_B}}{\binom{w_A}{w_B}} > \left(1 - \frac{\theta}{\sqrt{n}} \right)^{-\theta\sqrt{n}} \approx e^{\theta^2}.$$

If we choose θ large enough for

$$e^{\theta^2} > 2^{t+c},$$

then (d) follows from the resulting contradiction.   ∎

Theorem 12.   (a)   If C0–C4 and (2′), then there is a unique almost agreeing probability measure P given by

$$(\forall A_n) \, P(A_n) = 2^{-n}.$$

(b)   If in addition C7, then the almost agreeing P agrees with $\gtrsim$.

Proof.   (a)   Assume to the contrary that

$$(\exists n, A_n, B_n)\, A_n > B_n .$$

Then we construct an infinite sequence Z for which

$$(\forall m)\, A_n Z_m > B_n Z_m ,$$

thereby violating (2′). To recursively construct Z note that by the elementary properties of CP either $A_n 1 > B_n 1$ or $A_n 0 > B_n 0$. Whichever case is true let that then be Z_1. Hence, $A_n Z_1 > B_n Z_1$. Iterate this argument to generate

$$(\forall m)\, A_n Z_m > B_n Z_m .$$

Having reached a contradiction, we note that it can only be removed by

$$(\forall n, A_n, B_n)\, A_n \sim B_n .$$

Noting that for each n we have a 2^n-fold uniform partition of the sample space, we invoke Savage's theorem (Subsection IIC4) to obtain the existence of the unique almost agreeing probability P.

(b)   That the almost agreeing P in fact agrees with $\gtrsim$ satisfying C7 is an immediate corollary of Theorems II.3 and II.5. ∎

References

1. A. Kolmogorov, On Tables of Random Numbers, *Sankhyā Ser. A*, pp. 369–376, 1963.
2. D. Knuth, *The Art of Computer Programming*, Vol. 2, pp. 127–157. Reading, Massachusetts: Addison-Wesley, 1969.
3. P. Martin-Löf, Literature on V. Mises' Kollektivs Revisited, *Theoria* **35**, Pt. 1, 12–37, 1969.
4. A. Kolmogorov, Three Approaches to the Quantitative Definition of Information, *Problemy Peredači Informacii* **1**, 4–7, 1965.
5. D. Pager, On the Problem of Finding Minimal Programs for Tables, *Information and Control* **14**, 550–554, 1969.
6. H. Rogers, Jr., *Theory of Recursive Functions and Effective Computability*, pp. 1–24. New York: McGraw-Hill, 1967.
7. D. Loveland, A Variant of the Kolmogorov Concept of Complexity, *Information and Control* **15**, 510–526, 1969.
8. G. Chaitin, On the Length of Programs for Computing Finite Binary Sequences, *J. Assoc. Comput. Mach.* **16**, 145–159, 1969.
9. J. Hartmanis and J. Hopcroft, An Overview of the Theory of Computational Complexity, *J. Assoc. Comput. Mach.* **18**, 444–475, 1971.

10. G. Chaitin, Information-Theoretic Limitations of Formal Systems. Paper given at *Symp. Comput. Complexity, Courant Inst., New York Univ., New York, October 1971.*
11. J. Ziv, On the Complexity of an Individual Sequence, Elec. Eng. Pub. No. 146. Faculty of Elec. Eng., Technion, Haifa, Israel, June 1971.
12. E. Lehmann, *Testing Statistical Hypotheses*, pp. 60–63. New York: Wiley, 1959.
13. P. Martin-Löf, The Definition of Random Sequences, *Information and Control* **9**, 602–619, 1966.
14. P. Martin-Löf, Algorithms and Random Sequences. Ph.D. Thesis, Univ. of Erlangen, Germany, 1966.
15. A. Kolmogorov, Logical Basis for Information Theory and Probability Theory, *IEEE Trans. Information Theory* **IT-14**, 662–664, 1968.
16. G. Bruere-Dawson, *Pseudo-Random Sequences*, p. 44. M. S. Thesis, MIT, Cambridge, Massachusetts, June 1970.
17. P. Martin-Löf, On the Notion of Randomness, in *Intuitionism and Proof Theory* (A. Kino *et al.*, eds.), pp. 73–78. Amsterdam: North-Holland Publ., 1970.
18. C.-P. Schnorr, *Zufalligkeit und Wahrscheinlichkeit* (Lecture Notes of Math. 218). Berlin and New York: Springer-Verlag, 1971.
19. T. Fine, Stochastic Independence and Computational Complexity, *Proc. IFIP Congr. 1971*, Amsterdam: North-Holland, in press.
20. R. Solomonoff, A Formal Theory of Inductive Inference, Part I, *Information and Control* **7**, 1–22, 1964.
21. D. Willis, Computational Complexity and Probability Constructions, *J. Assoc. Comput. Mach.* **17**, 241–259, 1970.
22. R. Solomonoff, *Inductive Inference Research Status Spring 1967*. Rockford Res. Inst., July 1967, distributed by Clearinghouse, Dept. of Commerce.
23. J. Hintikka, Towards a Theory of Inductive Generalization, in *Logic, Methodology and Philosophy of Science* (Y. Bar-Hillel, ed.), p. 283. Amsterdam: North-Holland Publ., 1965.
24. G. Tartara, On Some Applications of Algorithmic Information Theory to Detection Problems, *Int. Symp. Information Theory, Noordwijk, Netherlands, June 15–19, 1970*, unpublished.
25. R. Solomonoff, A Formal Theory of Inductive Inference, Part II, *Information and Control* **7**, 224–254, 1964.

VI

Classical Probability and Its Renaissance

VIA. Introduction

The classical approach attempts to assess unique (prior) probabilities for random events even in the absence of extensive prior knowledge or information (e.g., extensive relative-frequency data or long data sequences) concerning the random experiment. In its early formulation by Laplace, through the principle of nonsufficient reason, equiprobable events were identified by the absence of reasons to expect the contrary — a balance of ignorance. Later rephrasing by Keynes as the principle of indifference or of sufficient reason avoided certain paradoxes by restricting the determination of equiprobability to cases where there was a balance of knowledge or information concerning the tendency for events to occur or propositions to be true.

Difficulties in assessing equiprobable events in complex experiments, problems with paradoxical conclusions, and questions of justification of the principle of indifference have led to axiomatic reformulations. The use of the principles of invariance and the information-theoretic principles of maximum entropy and mutual information seems to have enlarged the domain of classical probability to include unequal probability assignments. Furthermore, decision theory has provided some instances where a pragmatic justification for the classical approach can be developed. Yet, notwithstanding this progress, we are not optimistic about the value of the classical approach we now turn to outline.

VIB. Illustrations of the Classical Argument and Assignments of Equiprobability

The hallmark of the so-called classical or Laplacian approach to probability is the conversion of either complete ignorance or partial, symmetric knowledge concerning which of a set of alternatives is true, into a uniform probability distribution over the alternatives. The core of this approach is either the principle of nonsufficient reason [1] (assume alternatives to be equiprobable in the absence of known reasons to the contrary) or the principle of indifference [1] (assume alternatives to be equiprobable when there is a balance of evidence in favor of each alternative). It is presumed that we are able to recognize the equiprobable alternatives or events, whether they are as few as the two outcomes of the toss of a coin or as many as the values of a real parameter such as the probability of a sunrise tomorrow.

The following are examples typical of the domain of the classical approach to probability.

(1)   All outcomes are equally probable in the toss of a die.

(2)   The assumption of uniformly distributed phase in incoherent reception.

(3)   In a table of physical constants or a census report, all digits are equally probable as initial digits.

(4)   In a physical system of n identical particles, each capable of being in any of S states, all possible states of the system are equally probable, that is, each has probability n^{-S}.

(5)   It is as probable that an eggl is a gronk as that it is a zark.

(6)   Given that the sun has risen every day for N days, the probability that it will rise tomorrow is $(N + 1)/(N + 2)$.

(7)   The probability is 1/4 that the length of a chord, drawn at random inside a unit radius circle, will exceed $\sqrt{3}$ (the Bertrand paradox).

These examples provide the illustrative basis for our discussion of the role and status of appeals to the principle of indifference. Our conclusions will include the following: We cannot extract information (a probability distribution) from ignorance; the principle is ambiguous and applications often result in inconsistencies; our assessments of a balance of evidence need guidelines; and the classical approach to probability is neither an objective nor an empirical theory.

The first example provides a generic instance in which a supposedly clearly perceived symmetry in the physical die combines with the irrelevance of the face markings to yield the purportedly empirical conclusion that all faces are equally probable to appear. Interestingly enough a perusal of David [2] informs us that the ancient die was a bone from the heel of a sheep and quite irregular. It was only after many years of experimentation that the present-day cubical, symmetrical die evolved. We may speculate that this game of chance was developed, after much experience, to yield "easily evaluated" equiprobable outcomes; it is this lengthy experience that may be elliptically invoked rather than the principle of indifference. Another basis for our understanding of the behavior of die tossing, which does not involve the principle of indifference, is developed in connection with our second example.

In communications and detection theory we frequently use the model of incoherent reception—the phase of the signal incident at the receiver is a random variable uniformly distributed over a period of the carrier. Use of the classical approach would have us say that our ignorance of the phase of the received waveform is such that it is no more probable to be in the interval $(a, a + \epsilon)$ than in the interval $(b, b + \epsilon)$. Thus it must be uniformly distributed over a period. However, we can arrive at the same conclusion by means of another argument of the type introduced by Poincare [3]. We assume that the range from transmitter or target to the receiver is a random variable with some smooth density function $p(r)$. At, say, radar frequencies a period corresponds to a wavelength of about .1 meters. Typical ranges are of the order of 10^5 meters. Thus we may reasonably expect $p(r)$ to be approximately constant for changes in r of the order of a wavelength (1 part per 10^6). Since the phase is proportional to the range modulo the wavelength, it easily follows that under the above assumptions the density for the phase random variable should be approximately uniform over a period. This argument for a random phase requires no application of the principle of indifference.

The first two examples suggest that even where applicable the principle of indifference may be unnecessary. However, the third and fourth examples are ones for which there are few grounds short of indifference or experimentation. Recourse to the principle of indifference seems to indicate that the first digits in a table of physical constants or a census report are equiprobable. Unfortunately, experiment shows that the first digits are logarithmically, and not uniformly, distributed [4]!

The fourth example asserts that the system of physical particles obeys the Maxwell–Boltzmann statistics; so far as we know, all states are equally probable. However, we learn from experiment that

> Numerous attempts have been made to prove that physical particles behave in accordance with Maxwell–Boltzmann statistics, but modern theory has shown beyond doubt that this statistics does not apply to any known particles; ... (Feller[†] [5, p. 39]).

The physically realistic statistical mechanics models of Bose–Einstein or Fermi–Dirac in effect amount to different identifications of the equally probable cases than that which may have seemed apparent to a classical probabilist; for example in Bose–Einstein statistics we assume that the particles are indistinguishable and thereby combine the $n!$ states arising from permutations of the n particles into one alternative. This illustrates the dilemma encountered in applications of the principle of indifference when very little is known about a set of alternatives A. Ignorance concerning A is not easily distinguished from ignorance concerning a new set of alternatives A' derived from A either by combining elements of A or by subdividing them. An argument from ignorance is not an empirical method and cannot be expected to yield true empirical conclusions.

It is sometimes asserted that the classical approach is sensible even when we are ignorant of the nature of the alternatives. Perhaps this view has currency because we are often not fully ignorant about alternatives of interest to us. (How did they come to our attention?) The fifth example provides a fairly pure instance of ignorance concerning alternatives. It is beyond us what sense could be made of a probability assertion here. Haphazardly correct assumptions can only be expected to yield haphazardly correct conclusions.

The sixth example is attributed to Laplace. The statement was derived by assuming that sunrises are independent and identically distributed random events with an unknown probability p that a sunrise will occur on a given day. Laplace treats p as a random variable with a uniform density over $(0, 1)$, and derives the statement of the sixth example by use of Bayes theorem. Ignoring for the moment the grounds for the model, where did the uniform density come from? Laplace reasoned that since we are ignorant of the value of p, then given that $0 \leqslant a, b, a + \epsilon, b + \epsilon \leqslant 1$, it is as likely to be in the interval $(a, a + \epsilon)$ as in the interval $(b, b + \epsilon)$. If this were not the case, then we would be claiming a preferred location for p, and this represents knowledge we denied we had. Unfortunately, the same argument is applicable to some

---

[†] W. Feller, *Introduction to Probability Theory and Its Applications*, I, 2nd ed., p. 39. New York: Wiley, ©1957, by permission of John Wiley & Sons, Inc.

function of p, such as p^2, and a uniform density for p^2 is inconsistent with a uniform density for p.

The ambiguities inherent in an appeal to the principle of indifference can lead to inconsistent answers. If we are truly ignorant about a set of alternatives, then we are also ignorant about combinations of alternatives and about subdivisions of alternatives. However, the principle of indifference when applied to alternatives, or their combinations, or their subdivisions, yields different probability assignments. This criticism might not apply to the use of the principle of indifference based on a balance of substantial arguments in favor of each alternative.

The last example is a well-known one in geometric probability [6]. We can easily derive three contradictory answers to the probability of drawing a chord at random inside a unit radius circle that will be longer than $\sqrt{3}$. We may apply the principle of indifference either to the location of the midpoint of the chord, or to the end points of the chord lying on the circumference, or to the location of the midpoint of the chord given that it lies on some arbitrary diameter; each application results in a new answer. The details are available in the literature. This example reinforces our warnings that there are ambiguities in the selection of the set of alternatives which can lead to inconsistent results through use of the principle of indifference.

VIC. Axiomatic Formulations of the Classical Approach

1. The Principle of Invariance

It has been suggested, apparently first by Jeffreys [7] and then by others [8], that the principle of indifference can be placed on a sounder footing by interpretation through the use of the principle of invariance in much the same way as invariance is used in statistical decision theory [9]. Our prior knowledge concerning the probability distribution of an outcome in a random experiment is assumed to be representable in the form of a collection (usually a group) of transformations G. From each random experiment $\mathscr{E} = (\Omega, \mathscr{B}, P)$ we generate its orbit $\{\mathscr{E}_g, g \in G\}$ where

$$(\forall B \in \mathscr{B}) \, P_g(B) = P\big(g^{-1}(B)\big), \qquad g(\Omega) = \Omega, \qquad \mathscr{E}_g = (\Omega, \mathscr{B}, P_g).$$

The principle of invariance requires us to consider all of the random experiments in an orbit as equivalent and indistinguishable. Our prior knowledge does not enable us to discriminate between the $\{\mathscr{E}_g\}$ as to which best describes the actual experiment, and hence if there is to be a unique description, the $\{\mathscr{E}_g\}$ must be identical.

An example of an invariance argument is provided by the following revision of the Laplacian argument leading to a uniform prior density for the "unknown" p. Let the parameter p be represented as a point in a sample space Ω that is a circle of unit circumference,

$$p \leftrightarrow \omega = e^{i2\pi p} \in \Omega.$$

We wish to identify the prior distribution P on the usual Borel field of subsets of Ω which corresponds to having little prior knowledge of p. The group of transformations G which describes our prior knowledge, or the extent of our ignorance, is based on the one-parameter rotation group,

$$g(\omega) = e^{i\theta_g \omega}$$

(rotation of Ω through an angle of θ_g radius). On the basis of our prior knowledge we are unable to discriminate whether $\mathscr{E}$ or any $\mathscr{E}_g$ is a better description of the actual random experiment. Applying the principle of invariance, we require that these experiments be indistinguishable,

$$(\forall g)\, \mathscr{E}_g = \mathscr{E}.$$

It is easily verified that the unique P for which $P_g = P$ has the uniform density over Ω. Hence our prior knowledge is consistent only with a uniform prior distribution for p.

A source of difficulty with the invariance approach to classical probability is the necessity for the identification of the transformation groups that properly characterize our incomplete prior knowledge. For instance, in the sample presented above, why did we not enlarge G to include such transformations as $g(\omega) = \omega^2$? Had we done so, we would have found no solution for the invariant distribution P. On the one hand, a cautious selection of G may not lead to a unique prior distribution, thereby necessitating the introduction of additional principles to isolate the unique classical prior distribution. On the other hand, it is often the case that the use of invariance so overdetermines the prior distribution that there does not exist an invariant distribution.

A partial solution to the problem of overdetermination is to enlarge the class of prior "distributions" to include measures that are not prior distributions. For example, our prior knowledge may only be that the outcome of $\mathscr{E}$ is any real number x. The translation group

$$g(x) = x + t_g$$

might be employed to characterize this state of uncertainty. However, the only density function invariant under translation is the uniform density on $(-\infty, \infty)$ and it is not integrable as required by $P(\Omega) = 1$. Other applications of the principle of invariance given by Jaynes and Jeffreys also share this difficulty of seeming to yield nonintegrable prior densities. Hartigan [10] has presented other examples of invariance principles and evaluated the nonintegrable priors to which they lead. Such an approach of course can be objected to as not resulting in a theory of classical probability that agrees with the Kolmogorov axioms.

These observations suggest that the use of the principle of invariance is beset by many of the problems of the classical approach as well as by a few peculiar to itself. Even if these difficulties with the application of the principle of invariance could be mastered, there would still remain the problem of justifying the use of this principle and, indeed, of the classical approach.

2. Information-Theoretic Principles

The approach to classical probability through the information-theoretic principle of maximum entropy [8] (MEP) is a conservative one. The principle requires us to select as a prior distribution the least "informative" element of the family of distributions consistent with our prior knowledge. The prior knowledge is used only to specify a family $\mathscr{P}$ of possible prior distributions (e.g., we may know that we can restrict our attention to those distributions corresponding to integer-valued random variables of finite mean). Information not translatable into a restriction on the size of $\mathscr{P}$ (e.g., the result of a noisy observation of the mean) is ignored. If $\mathscr{P}'$ is a subset of the discrete distributions, then the MEP advises us to select as prior distribution P^* where

$$H(P) = -\sum_k P(X_k) \log P(X_k),$$

$$H(P^*) = \max_{P \in \mathscr{P}'} H(P);$$

$H(P)$ is the entropy of the discrete distribution P. There are difficulties in defining the MEP when $\mathscr{P}$ is not restricted to discrete distributions. While Tribus [11] attempts to approximate continuous distributions by discrete distributions assigning probability to the uniformly spaced values $\{k\delta\}$ for some small δ, Jaynes recognizes that the nondiscrete case cannot be so easily disposed of. Jaynes suggests that extension to the continuous case involves the determination of the density function that describes "complete ignorance," and he employs the principle of invariance to assist him in this questionable quest [8, p. 236].

As an example of the application of the MEP, we assume that our prior information is such that we can restrict our search for a classical prior to

$$\mathscr{P}' = \left\{ P : P(k) \geqslant 0, \sum_{k=1}^{\infty} P(k) = 1, \sum_{k=0}^{\infty} kP(k) = m \right\}.$$

Hence, we must

$$\max_{P \in \mathscr{P}'} - \sum_{k=1}^{\infty} P(k) \log P(k)$$

subject to

$$P(k) \geqslant 0, \qquad \sum_{k=1}^{\infty} P(k) = 1, \qquad \sum_{k=1}^{\infty} kP(k) = m.$$

The solution can be effected through the use of Lagrange multipliers, and is found to be

$$P(k) = \frac{1}{1 + m} \left(\frac{m}{1 + m} \right)^k, \qquad k = 0, 1, \dots .$$

The geometric distribution is suggested by the MEP as the classical prior for this problem.

A formal difficulty with the MEP becomes apparent if we modify the preceding example to

$$\mathscr{P}'' = \left\{ P : P(k) \geqslant 0, \sum_{k=0}^{\infty} P(k) = 1, \sum_{k=0}^{\infty} kP(k) < \infty \right\}.$$

There is now no distribution in $\mathscr{P}''$ that achieves maximum entropy. Another example that exploits the distinction between $\max_{P \in \mathscr{P}''} H(P)$ and $\sup_{P \in \mathscr{P}''} H(P)$ is the following:

$$\mathscr{P}''' = \left\{ P : (\forall P)(\exists N) \sum_{k=0}^{N} P(k) = 1, \sum_{k=0}^{\infty} kP(k) = 1 \right\}.$$

It follows from the preceding example that there is no maximizing P^* but there is a distribution P' for which the supremum is achieved,

$$P'(k) = 2^{-k-1} \qquad \text{if} \quad k = 0, 1, \dots .$$

Thus there are formal difficulties with the application of the MEP even to discrete distributions.

A generalization of the MEP to a maximum mutual information principle (MMIP) has been recently suggested by Kashyap [12]. The advantage claimed for the MMIP is that it can incorporate information concerning the prior distribution derived from the observation of related random variables. Such information would not, in general, reduce $\mathscr{P}$ and would have been ignored in applications of the MEP, although it appears to be relevant to the choice of a classical prior. Let p be the parameter for which we seek a prior distribution P. On the basis of prior information we can restrict P to a set $\mathscr{P}'$. Let X be a random variable related to p through the conditional distribution $F_{X|p}$. Assuming for convenience that P has a density f_p and $F_{X|p}$ has a density $f_{X|p}$, then we can define the mutual information $I(X; p)$ between X and p through

$$I(X; p) = \iint \log \left[\frac{f_{X|p}}{f_X} \right] f_{X|p} f_p \, dx \, dp,$$

where

$$f_X = \int f_{X|p} f_p \, dp.$$

The MMIP suggests adopting as prior the density f_p that achieves $\max_{P \in \mathscr{P}'} I(X; p)$.

As might be anticipated, there are as yet unresolved formal difficulties with the application of the MMIP. In addition to counterparts of the difficulties with the MEP we have to eliminate such nuisances as X happening to be independent of p; independence of X and p forces $I = 0$ without regard to P.

VID. Justifying the Classical Approach and Its Axiomatic Reformulations

1. Classical Probability and Decision under Uncertainty

It is possible to justify pragmatically to some degree this penchant of classical probability to assign equal probability to the occurrence of events when our prior knowledge is balanced with respect to their occurrences. Chernoff [13] and Milnor [14] have both proposed axiomatizations of rational decision-making under uncertainty which appear to justify the classical approach. Although we discuss the nature of a decision problem somewhat more fully in Chapter VIII, for the present we may oversimplify and represent it by a matrix $R = [R_{ij}]$ where R_{ij} is the numerical loss sustained if we employ decision rule, or take act, d_i when s_j is the true state of nature. We assume that the set $\mathscr{S}$ of states of nature is finite. The decision problem is then the selection of a best decision rule or row in the matrix R.

Milnor has presented the following axioms to describe one type of "rational" preferencing $\gtrsim$ between decision rules.

M1. The binary relation $\gtrsim$ is a complete (linear) ordering of decision rules (rows).

M2. $\gtrsim$ is invariant under a permutation of rows or columns.

M3. $(\forall j) \, (R_{ij} > R_{kj}) \Rightarrow d_i > d_k$ (strict preference).

M4. The ordering between two decision rules is unchanged by the introduction of a new decision rule.

M5. The ordering is unchanged if R is replaced by $R_k = [R_{ij} + c\delta_{ik}]$ for any k.

Milnor then verifies that any ordering $\gtrsim$ satisfying M1–M5 can be represented as follows:

$$d_i \gtrsim d_j \Leftrightarrow \frac{1}{n} \sum_{k=1}^{n} R_{ik} \leqslant \frac{1}{n} \sum_{k=1}^{n} R_{jk} \, .$$

This result is easily interpreted in terms of the Bayes criterion of preferring that decision rule yielding the minimum expected loss. The expression $(1/n) \sum_{k=1}^{n} R_{ik}$ is the expected loss incurred by using rule d_i when all of the states of nature $\{s_1, ..., s_n\}$ are equally probable.

Insofar as axioms M1–M5 are deemed acceptable when we have a balance of arguments in favor of the occurrence of each state of nature, they supply a pragmatic justification for the equiprobability assignment suggested by classical probability.

2. Principle of Invariance

The thrust behind the principle of invariance seems to derive from the impossible task that would confront the classical probabilist were it to fail to apply. If, on the basis of our limited prior knowledge, we are unable to discriminate between elements of a set of distinct probability models $\{\mathscr{E}_g, g \in G\}$, then it is impossible to assign a unique prior distribution. Hence the necessity for a principle of invariance to assure us that $\{\mathscr{E}_g, g \in G\}$ are equivalent descriptions of the actual random experiment.

Of course, while the principle of invariance may be required to ensure the existence of an answer to the question of a prior distribution, its application in no way assures us that the question itself is not ridiculous. Furthermore, it must be noted that the use of the principle of invariance in statistical decision theory has occasionally led to unacceptable results. There are nonpathological examples of decision problems for which the invariant decision rules are inadmissible [15]. This possibility suggests caution in wide-ranging applications of the principle. Finally, we need a better understanding of the apparently unstated procedure by which we convert our prior knowledge into the appropriate collection of transformations. Is the conversion to be subjective or objective? Is the collection necessarily a group?

3. Information-Theoretic Principles

Jaynes[†] has argued in defense of the MEP in its relation to relative-frequency theories that

> ... we are entitled to claim that probabilities calculated by maximum entropy have just as much and just as little correspondence with frequencies as those calculated from any other principle of probability theory [8, p. 233].

In view of our discussion of frequency theories in Chapter IV, we do not find this defense especially encouraging. Nor is it clear that Jaynes has established this point.

A better line of defense for the information-theoretic principles has been suggested by Kashyap [12]. Kashyap attempts a pragmatic justification for the MMIP by proving that under certain conditions the application of the MMIP leads to a good strategy in that the derived prior is the least favorable prior [9, p. 34] for a particular game. However, the few results thus far available in this area preclude a judgment as to the potential strength of this defense.

VIE. Conclusions

We remain unconvinced that the approaches of classical probability, in either its original or axiomatic reformulations, are sufficiently free from ambiguity in their application, treat a wide enough domain of states of prior knowledge, or have been adequately justified. The interpretation

---

[†] E. Jaynes, Prior Probability, *IEEE Trans. Systems Sci. Cybernetics* **SSC-4**, 233, 1968.

of a classical prior distribution is particularly problematic: it seems to be an objectification of subjective prior knowledge. Extracting unique quantitative probabilities from ignorance or very little prior knowledge can only harmfully obscure our ignorance. At the least there should be a distinction between a prior distribution arrived at on the basis of extensive knowledge (e.g., the order of cards in a deck after I have just thoroughly shuffled the deck) and the same prior resulting from the legerdemain of classical probability (e.g., the order of cards in a deck produced by a stranger).

If the alternatives, about whose occurrence we are either completely ignorant or have a balance of arguments in favor of each one, are not equiprobable, then what are they? This dilemma arises from an implicit assumption that probabilities can always be assigned. The assumption of the universal applicability of probability is denied in modern statistical theory. Statisticians, other than Bayesians, admit a distinction between random and unknown parameters; there are no probabilities assigned to the latter. The resolution to the dilemma may lie in the realization that the assertion that complete ignorance, or even a balance of arguments, is not consistent with an unequal assignment of probabilities is also compatible with the nonexistence of probabilities. I think it wiser to avoid the use of a probability model when we do not have the necessary data than to fill in the gaps arbitrarily; arbitrary assumptions yield arbitrary conclusions.

References

1. J. M. Keynes, *A Treatise on Probability*, Chapter IV. New York: Harper, reprinted 1957.
2. F. N. David, *Games, Gods and Gambling*, Chapter I. London: Griffin, 1962.
3. H. Reichenbach, *The Theory of Probability*, pp. 355–359. Berkeley: Univ. of California Press, 1949.
4. W. Feller, *Introduction to Probability Theory and Its Applications, II*, pp. 62–63. New York: Wiley, 1966.
5. W. Feller, *Introduction to Probability Theory and Its Applications, I*, 2nd ed., p. 39. New York: Wiley, 1957.
6. B. Gnedenko, *The Theory of Probability*, 4th ed., p. 47. Bronx, New York: Chelsea, 1967.
7. H. Jeffreys, *Theory of Probability*. London and New York: Oxford Univ. Press (Clarendon), 1948.
8. E. Jaynes, Prior Probability, *IEEE Trans. Systems Sci. Cybernetics* **SSC-4**, 227–241, 1968.
9. T. Ferguson, *Mathematical Statistics*, Chapter 4. New York: Academic Press, 1967.
10. J. Hartigan, Invariant Prior Distributions, *Ann. Math. Statist.* **35**, 836–845, 1964.

11. M. Tribus, *Rational Descriptions, Decisions and Designs*, p. 130. Oxford: Pergamon, 1969.

12. R. Kashyap, Prior Probability and Uncertainty, *IEEE Trans. Information Theory* **IT-17**, 641–650, 1971.

13. H. Chernoff, Rational Selection of Decision Functions, *Econometrica* **22**, 422–443, 1954.

14. J. Milnor, Games Against Nature, in *Decision Processes* (R. Thrall, C. Coombs, and R. Davis, eds.), pp. 49–59. New York: Wiley, 1954.

15. C. Stein, Inadmissibility of the Usual Estimator for the Mean of a Multivariate Normal Distribution, *Proc. Symp. Math. Statist. and Probability 3rd, Berkeley* **1**, pp. 197–206. Berkeley: Univ. of California Press, 1956.

VII

Logical (Conditional) Probability

VIIA. Introduction

In the preceding chapter we encountered the view that the probability of an event might be assessed from a balance of evidence, or lack of evidence, in favor of the occurrences of each of an exhaustive set of mutually exclusive alternatives, some subset of which comprises the event. However, the classical theory lacked guidelines for the identification of a balance of evidence and was open to inconsistent application. Furthermore, the domain of the classical theory, although not the domains of its reformulations, covered only those situations where the alternatives were of equal weight; only equiprobability assignments were possible. Spurred by the goals of classical probability and its failure to achieve them, there has been a development, by Keynes [1], Jeffreys [2], Koopman [3], Carnap [4], and others, of more flexible theories of logical probability. These theories of inductive logic or nondemonstrative reasoning aim at providing rational or reasonable assessments of the degree to which an hypothesis H is supported by evidence E or, more loosely, the probability that H is true given E.

The formal domain of a theory of logical probability is generally a set of inferences between statements or propositions in a language, rather than the set of statements themselves, and it is distinct from the domain of an empirical theory (set of events or experimental outcomes) and the domain of a subjective theory (set of beliefs of an individual). Formally the distinction between an inference-domain theory and one with a statement-domain is parallel to that between conditional and absolute probability. In general, we can generate a statement-domain theory from an inference-domain theory by restricting the latter to the case of a fixed, reasonably selected, evidence proposition (e.g., a tautology or logical truth). However, the generation of an inference-domain theory from a statement-domain theory is likely to be more difficult (see Subsection IIE3).

Logical probability attempts to explicate induction by defining a logical relation between an evidence statement and an hypothesis statement that is a generalization of the relations of implication and contradiction available from deductive logic. There are several views, however, as to the formal and interpretive nature of the evidence–hypothesis relation.

As was mentioned in Chapter I, there is a weak formal concept of classificatory probability. While ignored by other approaches to probability, the classificatory concept has some interest when examined from the viewpoints of modal logic and logical probability. We then examine Koopman's somewhat stronger theory of comparative logical probability. While Koopman discusses an agreeing quantitative logical probability theory, his subjective interpretation of probability gives us few guidelines for the actual determination of logical probability. Finally, we consider Carnap's more ambitious program for quantitative logical probability. Carnap originally sought a unique, quantitative logical relation, called degree of confirmation (d.c.), to measure the support one statement lends to another. This support is seen as being of an analytical, necessary, or logical nature rather than of a synthetic, contingent, or empirical nature. Just as implication is determinable from the meaning of statements irrespective of their truth, so should a degree of confirmation be determined without regard to either the truth of the statements involved or the contingent, factual aspects of the world. While it appears that Carnap's theory is as yet incomplete, it is sufficiently developed for us to give examples of its application.

The various theories of logical probability have in common, beyond agreement as to the domain of probability properly being that of inferences, the derivation of quantitative probability by guarded use of the principle of indifference and the desire to justify themselves by demonstrating some form of agreement with a relative-frequency outlook. In Carnap's and Koopman's theories, logical probability is either an estimate of or converges to the relative-frequency of the number of individuals for which the hypothesis statement proved to be correct (truth frequency). Going further, Carnap insists that logical probability also be pragmatically justifiable by appeal to a theory of rational decision making. Correct reasoning is to find its *raison d'être* in correct or rational decisions and judgments.

VIIB. Classificatory Probability and Modal Logic

While the classificatory concept of probability ("*A* is probable") is a very weak characterization of randomness or uncertainty in many interpretations of probability and has therefore been ignored, it finds a place in logical probability through modal logic [5, 6]. Paralleling common usage and notwithstanding our earlier remarks, we will treat a version of classificatory logical probability in which the domain is the set of propositions rather than the set of inferences between propositions. Although we will not do so, the ensuing discussion can be formally applied to inferences.

In brief, a modal probabilistic logic for propositions adjoins to a propositional logic the operator "$\mathscr{P}$" which when prefixed to a proposition p, "$\mathscr{P}p$," is read either as "probably p" or "p is probably true." The operator $\mathscr{P}$ is a mapping from the space of propositions into the space of propositions. The relation between $\mathscr{P}$ and the usual propositional logic symbols of conjunction ($\wedge$), disjunction ($\vee$), negation ($\sim$), implication ($\Rightarrow$), and parenthesis [(,)], can be axiomatized in several, inequivalent, ways. A reasonable set of axioms for $\mathscr{P}$ might include the following:

M1. $(\forall p)\, (\mathscr{P}(p \vee \sim p))$.

M2. $(\forall p) \sim (\mathscr{P}p \wedge \mathscr{P}(\sim p))$.

M3. $(\forall p, q)\, ((p \Rightarrow q) \wedge \mathscr{P}p \Rightarrow \mathscr{P}q)$.

Axiom M1 asserts that a certain kind of tautology is probable. In an event language model in which the propositions describe the occurrences of events, the counterpart of M1 is that the occurrence of the certain event (Ω) is probable. Axioms M1 and M3 together assert the probability of any tautology. Axiom M2 denies the existence of a proposition p for which p and $\sim p$ are both probable. In terms of the usual quantitative probability P, M2 could be modeled for a proposition p_A describing an event A by

$$(\exists t \geqslant \tfrac{1}{2}) \text{ “}\mathscr{P}p_A\text{”} \quad \text{iff} \quad \text{“}P(A) > t.\text{”}$$

Axiom M3 is an axiom of detachment. In the event model M3 corresponds to "B is probable and $A \supseteq B$ implies A is probable." While Hempel [7] has challenged the acceptability of detachment rules in probability, we will accept M3.

Another possible axiom is

M4. $(\forall p) \, (\mathscr{P}p \vee \mathscr{P}(\sim p)).$

Axiom M4 asserts that for any proposition, either it or its contradictory is in the domain of the operator $\mathscr{P}$. However, it might be felt that the implications of M2 and M4 are too strong. In particular, M2 and M4 imply that

$$(\forall p) \, (\mathscr{P}p \vee \sim\mathscr{P}p);$$

that is, for every proposition we can determine whether or not it is probable. Some propositions may be neither probable nor improbable.

In modal logic we are also concerned with the formal properties of iterated modalities, a simple example being "$\mathscr{P}\mathscr{P}p$" (read "probably p is probable"). A possible basis for interpreting the iterated modality of probability might be the following analogy with parametric statistical models, wherein we do not assume M4. Assume that there is a family $\{P_\theta \,, \theta \in \varTheta\}$ of probability measures, one of which correctly describes the random experiment, and that there is a prior distribution π on a suitable σ-field of subsets of $\varTheta$ that includes singleton events. Introduce the correspondences:

Classificatory logical probability Proposition "p_A"	*Statistical* Event "A" described by "p_A"
$\mathscr{P}p_A$	$\displaystyle\sum_{\theta \in \varTheta} \pi(\theta) \, P_\theta(A) > t \geqslant \tfrac{1}{2}$
$\mathscr{P}\mathscr{P}p_A$	$\pi(\{\theta : P_\theta(A) > t\}) > t$

With this interpretation, suitable for the commonly encountered problems of parametric estimation and decision-making, we see that the reduction of iterated modalities is contingent upon $\varTheta$ and π and cannot be axiomatized. Hence we cannot assert, for example,

$$\mathscr{P}p \Rightarrow \mathscr{P}\mathscr{P}p \quad \text{or} \quad \mathscr{P}\mathscr{P}p \Rightarrow \mathscr{P}p;$$

the truth of these implications depends on extralogical considerations.

The preceding formulation of the modal operator "$\mathscr{P}$" is quite different from any of the several formulations of the modalities of possibility "M" or necessity "N." While

$$(\forall p)\,(p \Rightarrow Mp),$$

it is not true that

$$(\forall p)\,(p \Rightarrow \mathscr{P}p);$$

an event can have occurred without its having been probable to occur. Furthermore, a consequence of M1–M4 is

$$(\forall p)\,(\mathscr{P}p \Leftrightarrow \sim\!\mathscr{P}(\sim\!p)).$$

This conflicts with the usually assumed relation

$$(\forall p)\,(Mp \Leftrightarrow \sim\!N(\sim\!p)),$$

unless we make the strange identification $M \equiv N$. Hence, "probability" and "possibility" are very different modalities.

VIIC. Koopman's Theory of Comparative Logical Probability

1. Structure of Comparative Probability

Koopman's theory of logical probability [3] applies to inferences between what he, informally, designates as experimental propositions (statements about events whose truth or falsity is determinable from the performance of an experiment). He introduces a quaternary conditional comparative probability relation, involving two evidence statements, E_1 and E_2, and two hypotheses, H_1 and H_2, that is written "$H_1/E_1 \precsim H_2/E_2$" and read "H_1 on E_1 is no more probable than H_2 on E_2." It is presumed that $\precsim$ is prescribed by an individual, based on his intuition, subject to rationality properties established in nine axioms. With respect to the origin of the $\precsim$ relation, Koopman has said,

> . . . the authority for the first ($\precsim$) proposition does not reside in any general law of probability, logic, or experimental science. And the notion presents itself that such primary and irreducible assumptions are grounded on a basis as much of the aesthetic as of the logical order[†] [3, p. 774].

[†] Reprinted with permission of the publisher, The American Mathematical Society, from *Bulletin of the American Mathematical Society*, Copyright © 1940, Vol. 46, p. 774.

Koopman's axioms for $\lesssim$ are as follows. The axioms connecting $\lesssim$ to logical implication are

K1. $(E_2 \Rightarrow H_2) \Rightarrow (H_1/E_1 \lesssim H_2/E_2)$.

K2. $(E_1 \Rightarrow H_1) \wedge (H_1/E_1 \lesssim H_2/E_2) \Rightarrow (E_2 \Rightarrow H_2)$.

The axioms extending the $\lesssim$ relation, or imposing constraints on permitted assignments, include

K3. *(Reflexivity)* $H/E \lesssim H/E$.

K4. *(Transitivity)* $(H_1/E_1 \lesssim H_2/E_2) \wedge (H_2/E_2 \lesssim H_3/E_3)$
$\Rightarrow (H_1/E_1 \lesssim H_3/E_3)$.

K5. $(H_1/E_1 \lesssim H_2/E_2) \Rightarrow (\sim H_2/E_2 \lesssim \sim H_1/E_1)$.

The next two axioms essentially state the product rule for conditional probability.

K6a. Let S_1 and S_2 be non-self-contradictory statements and $(S_1 \Rightarrow S_1' \Rightarrow S_1'')$, $(S_2 \Rightarrow S_2' \Rightarrow S_2'')$. Then

$$(S_1/S_1' \lesssim S_2/S_2') \wedge (S_1'/S_1'' \lesssim S_2'/S_2'') \Rightarrow (S_1/S_1'' \lesssim S_2/S_2'').$$

K6b. $(S_1/S_1' \lesssim S_2'/S_2'') \wedge (S_1'/S_1'' \lesssim S_2/S_2') \Rightarrow (S_1/S_1'' \lesssim S_2/S_2'')$.

K7. Assume the notation of K6 and $S_1/S_1'' \lesssim S_2/S_2''$. If either symbol in $(S_2/S_2', S_2'/S_2'')$ has the relation $\lesssim$ to either symbol in $(S_1/S_1', S_1'/S_1'')$, then the remaining symbol in the second set has the relation $\lesssim$ to the remaining symbol in the first set.

The remaining two axioms, omitted here, become theorems if the ordering of $\lesssim$ is complete; that is, if for all E_1, E_2, H_1, and H_2 either $H_1/E_1 \lesssim H_2/E_2$ or $H_2/E_2 \lesssim H_1/E_1$. Koopman, of course, does not require that all H/E pairs be comparable by $\lesssim$; even comparative probability is not assumed to be universally applicable.

Koopman's axioms for the conditional comparative probability of hypothesis statements given evidence statements have parallels with the axioms presented in Section IIE for the conditional comparative probability of events. Curiously, K2 implies that the only event at least as probable as the certain event is the certain event or equivalently, that null-equivalent events must be null. We did not make this assumption, nor is it usual in probability theory. Interestingly though in the complexity approach to probability it appears that a zero probability event never occurs [8].

2. Relation to Conditional Quantitative Probability

The transition from a "subjectively" selected comparative relation $\lesssim$ to a quantitative probability assignment for H/E is based on the postulated existence of an infinite sequence of "n-scales."

Definition. A set $\{S_1, ..., S_n\}$ is an n-scale if and only if:

(1) At least one S_i is non-self-contradictory.
(2) The conjunction $S_i \wedge S_j$ (to be read "S_i and S_j") of any two distinct statements is a contradiction.
(3) If E is the logical disjunction of the $\{S_i\}$,

$$E = \bigvee_{i=1}^{n} S_i,$$

where $\bigvee_{i=1}^{n} S_i$ is read as "S_1 or S_2 or ... or S_n," then

$$(\forall i, j)\ S_i/E \lesssim S_j/E.$$

Koopman's axiom is

K8. $(\forall n)\ \exists\{S_i\}$ an n-scale.

Axiom K8 can be recognized as a stronger form of Savage's almost uniform partition hypothesis (Subsection IIC4). As we would expect from the discussion of Savage's axiom, K8 only assures us of the existence of a finitely additive, almost agreeing probability P; that is,

$$(H/E \lesssim H'/E') \Rightarrow P(H/E) \leqslant P(H'/E'),$$

but the reverse implication is not necessarily true. Nor does it follow from K8 that $P(H/E)$ has the usual property of conditional probability that

$$P(H/E \wedge E') = \frac{P(H \wedge E'/E)}{P(E'/E)}.$$

Insight into the existence of a quantitative theory is available from the analyses of Sections IIC and IIE, and we will not reinvestigate this question here.

3. Relation to Relative-Frequency

A relation between Koopman's quantitative logical probability P and relative-frequency can be drawn as follows. Consider a language describing repeated experiments in which $\{a_i\}$ represent the trials, Sa_i is the statement of a "success" on trial i, and 0_n is the integer corresponding to the number of successful trials in the first n repetitions. Koopman [9, p. 185] establishes

Theorem 1. Let the evidence E be such that

(1) $(\forall n)(\forall \{i_j\}, \{k_j\}) \left(\left(\bigvee_{j=1}^{n} Sa_{i_j}/E \right) \lesssim \left(\bigvee_{j=1}^{n} Sa_{k_j}/E \right) \right)$;

(2) $\lim_{n \to \infty}(0_n/n) = p$.

Then
$$P(Sa_i/E) = p.$$

The evidence E must be such that the order of the trials is irrelevant (permutation invariance); the logical probability of the events of at least one success in n trials does not depend on which n trials we examine. The theorem also assumes that the limit of the relative-frequency of success actually exists, is known, and stated in E. The permutation invariance hypothesis is just de Finetti's hypothesis of exchangeable events (see Sections IVB and VIIE).

4. Conclusions

The significance of Theorem 1 is not the justification of the use of Koopman's logical probability as the estimate of "true" relative-frequency but rather a verification of the reasonableness of the proposed definition of quantitative logical probability. It would be surprising if under the stringent hypotheses of the theorem we found that $P(Sa_i/E) \neq p$. As for the determination of probability directly from relative-frequency, Koopman is well aware that a conclusion asserting a probability relation must be preceded by an antecedent assumption involving a probability relation.

Koopman's analysis of comparative and quantitative probability leaves to the intuition of the user the selection of the comparative relation and a suitable sequence of n-scales. He does not believe it possible to supply compulsory instructions for either selection and is content to leave matters with an axiomatic system capable of supporting an infinite variety of comparative assignments.

VIID. Carnap's Theory of Logical Probability

1. Introduction

In Carnap's view the theory of logical probability concerns a quantitative relation $C(H/E)$ between an hypothesis H and evidence E, called the degree of confirmation function (d.c.), that expresses the degree to which E implies or supports H. Inductive reasoning or inference is a process of assessing the degrees of confirmation of various hypotheses given an evidence statement rather than a process of choosing the uniquely correct or true hypothesis. The choice of a rational or "best" d.c. is to be made on analytic or logical grounds rather than on synthetic or empirical grounds and is to be independent of contingent facts. In this, induction parallels deduction, wherein determinations of implication or contradiction are made without regard to the contingent truth of the statements. All empirical or synthetic knowledge is presumed included in the evidence E. We reserve to Section VIIG our doubts as to the meaningfulness of this analytic/synthetic dichotomy.

Carnap structures the confirmation function C through three sets of axioms [10].† The first set of axioms relates to strict coherence and arises from a desire to develop C so that it will be useful for the purpose of making rational decisions. The second set of axioms concerns invariance properties for C and is motivated by the desires that there be no *a priori* distinguished predicates or individuals and that the relation between propositions H and E depends only on the subset of predicates and individuals they reference. The third set of axioms is concerned with ensuring the ability to learn from experience; they prescribe the behavior of C in certain "intuitively clear" instances of inductive inference.

The sort of languages $\mathscr{L}$ for which Carnap was able to develop his theory of logical probability most fully have finitely many individuals $\{a_i\}$ and finitely many families $\{\mathscr{P}_i\}$ of one-place "primitive" predicates. By a family $\mathscr{P}_i$ of predicates we mean a finite collection $\{P_{i,j}\}$ of predicates such that each individual is describable by one and only one predicate in each family. Other terms and notations of interest to us are as follows.

Definition. A complete description of an individual is a predicate Q of the form

$$Q = \bigwedge_i P_{i,j_i}$$

where the conjunction is over all families.

---

† Carnap's final remarks on logical probability appear to be those in [22].

The set of all complete descriptions of an individual will be denoted by $\mathscr{D}$, and q will denote the number of elements in $\mathscr{D}$. If there are K families of predicates and $\mathscr{P}_i$ has n_i elements, then

$$q = \prod_{i=1}^{K} n_i \, .$$

Definition. A state description S is a proposition of the form

$$S = \bigwedge_j Q_{i_j} a_j \, ,$$

where the conjunction is over all individuals.

A state description is a complete description of the world consisting of all the individuals $\{a_i\}$. The set of all state descriptions will be denoted by $\mathscr{S}$.

Definition. The range $\mathscr{R}(H)$ of a proposition H is given by

$$\mathscr{R}(H) = \{S : S \in \mathscr{S}, S \Rightarrow H\}.$$

For example, we may take N individuals $\{a_i\}$ to refer to the N repetitions of an experiment such as the toss of a die, and assume one family $\mathscr{P} = \{P_i\}$ of predicates, where "$P_i a_j$" means "i spots appeared on trial j." When there is only one family of predicates each P_i is itself a complete description. A state description S would be of the form "$\bigwedge_{j=1}^{N} P_{i_j} a_j$" meaning "$i_1$ spots appeared on trial 1 and ... and i_N spots appeared on trial N". The range of, say, $P_j a_k$ is given by

$$\mathscr{R}(P_j a_k) = \left\{ S : (\exists j_1, ..., j_{k-1}, j_{k+1}, ..., j_N) \left(S = P_j a_k \wedge \bigwedge_{n \neq k} P_{j_n} a_n \right) \right.$$

and contains 6^{N-1} state descriptions.

After presenting and examining Carnap's axioms for logical probability or d.c., we consider the relation of d.c. to relative-frequency phenomena, illustrate possible applications for d.c., and proceed to a brief critique of d.c.

2. Compatibility with Rational Decision-Making

In his later writings Carnap grew more insistent on the point that the formal analytic theory of d.c. be compatible with applications to rational decision-making. To relate d.c. to decision-making we introduce the notion of a gamble g on a set $\{H_i\}$ of mutually contradictory statements $((\forall i \neq j) \sim (H_i \wedge H_j))$; g assigns a numerical payoff $g(H_i)$ to each H_i. The d.c. $C(H/E)$ is to have the role of evaluating gambles through the "expected value"

$$v(g) = \sum_i C(H_i/E)\, g(H_i \wedge E)$$

when

$$E \Rightarrow \bigvee_i H_i, \qquad (\forall i \neq j)\, E \Rightarrow \sim(H_i \wedge H_j).$$

Gamble g_1 is to be as good as g_2 if and only if $v(g_1) \geqslant v(g_2)$. This evaluation procedure would be more compelling if g is a utility function [11], but it is not. The *status quo* is taken as a gamble with zero payoffs. Hence, g is as good as the *status quo* if and only if $v(g) \geqslant 0$.

The requirement of strict coherency arises from the rationality of rejecting gambles such that given E you can never achieve a positive payoff but can sustain a negative payoff. Formally we require of the d.c. that to be useful in decision-making it satisfy

L1. *(Strict coherence)* $\Big(E \Rightarrow \bigvee_i H_i, (\forall i \neq j)\, E \Rightarrow \sim(H_i \wedge H_j),$

$(\forall i)\, g(H_i \wedge E) \leqslant 0, (\exists j)\, g(H_j \wedge E) < 0 \Big) \Rightarrow v(g) < 0.$

A characterization of the class of strictly coherent d.c. is given by

Theorem 2 (de Finetti–Kemeny [12]). C is strictly coherent iff (throughout E, E', and E'' are not self-contradictory):

(1) $0 \leqslant C(H/E) < \infty.$
(2) $(H \Leftrightarrow H', E \Leftrightarrow E') \Rightarrow C(H/E) = C(H'/E').$
(3) $((E \Rightarrow H), (E' \not\Rightarrow H')) \Rightarrow (C(H/E) > C(H'/E')).$
(4) $(E \Rightarrow \sim(H \wedge H')) \Rightarrow (C(H \vee H'/E) = C(H/E) + C(H'/E)).$
(5) $C(H \wedge E'/E) = C(H/E \wedge E')\, C(E'/E).$

Hence $C(H/E)$ has the basic formal properties of a finitely additive conditional probability providing that we adopt the convention that

$$(E \Rightarrow H) \Rightarrow C(H/E) = 1,$$

and remember that

$$C(H/E) = 1 \Rightarrow (E \Rightarrow H).$$

Furthermore, if we select a tautology (logical truth) T and define

$$m(H) = C(H/T),$$

then we can represent

$$C(H/E) = \frac{m(H \wedge E)}{m(E)}.$$

The function m has the formal properties of a finitely additive probability measure with the added constraint that

$$m(H) = 1 \Rightarrow H \text{ a tautology.}$$

In terms of the language $\mathscr{L}$ described in Subsection VIID1, we see that the confirmation C can be represented by

$$C(H/E) = \frac{\sum_{S \in \mathscr{R}(H \wedge E)} m(S)}{\sum_{S \in \mathscr{R}(E)} m(S)}.$$

The strictly coherent confirmation functions, called regular by Carnap, satisfy the constraint

$$(\forall S \in \mathscr{S})\,(m(S) > 0).$$

If we wish to extend $\mathscr{L}$ to include countably many individuals, then we have to relax the requirement of strict coherence to that of coherence:

$$(\forall S \in \mathscr{S})\,(m(S) \geqslant 0), \qquad C(H/E) = \frac{\sum_{S \in \mathscr{R}(H \wedge E)} m(S)}{\sum_{S \in \mathscr{R}(E)} m(S)}.$$

3. Axioms of Invariance

In his attempts to narrow the class of confirmation functions to a unique "best" confirmation function, Carnap [10] invoked several axioms of invariance much as is done in classical probability. In terms of the language $\mathscr{L}$ we may state Carnap's axioms as

L2. $C(H/E)$ is invariant under any permutation of the individuals $\{a_i\}$.

L3. $C(H/E)$ is invariant if the set of individuals $\{a_i\}$ is augmented provided that no quantifiers ($\forall, \exists$) appear in either H or E.

L4. $C(H/E)$ is invariant under any permutation of the predicates $\{P_{ij}\}$ within a family $\mathscr{P}_i$.

L5. $C(H/E)$ is invariant under any permutation of two families $\mathscr{P}_i$ and $\mathscr{P}_j$ provided that $n_i = n_j$.

A strengthened version of L4 and L5 is

L6. $C(H/E)$ is invariant under any permutation of the complete descriptions in $\mathscr{D}$.

Finally, Carnap suggests

L7. $C(H/E)$ is invariant under augmentation of the set $\{\mathscr{P}_i\}$ of families of predicates.

The axioms apply as well of course to $m(H) = C(H/T)$.

To understand the implications of these invariance axioms it suffices, since C is determined by m and m is finitely additive, to characterize $m(\wedge_{j=1}^{n} Q_{i_j} a_j)$ where the conjunction $\wedge_{j=1}^{n} Q_{i_j} a_j$ would be a state description if $\mathscr{L}$ had only n individuals. A first result is given by

Lemma 1. If C satisfies L1–L3, then there is a unique measure F such that

$$R_q = \left\{ (p_1, ..., p_q) : p_i \geqslant 0, \sum_{i=1}^{q} p_i = 1 \right\},$$

$$F(R_q) = 1,$$

$$m\left(\bigwedge_{j=1}^{n} Q_{i_j} a_j \right) = \int \cdots \int \left(\prod_{j=1}^{n} p_{i_j} \right) F(dp_1, ..., dp_q),$$

and q is the number of complete descriptions.

Proof. All proofs of results can be found in the Appendix to this chapter. ∎

Equivalently, we may write that under L1–L3

$$m\left(\bigwedge_{i=1}^{n} Q_{i_j} a_j \right) = \int \cdots \int \left(\prod_{i=1}^{q} p_i^{0_i} \right) F(dp_1, ..., dp_q), \qquad (*)$$

where 0_i is the number of occurrences of Q_i in $\{Q_{i_j}\}$;

$$0_i \geqslant 0, \qquad \sum_{i=1}^{q} 0_i = n.$$

Given Lemma 1, we can directly, but not too fruitfully, explore the consequences of adjoining L4 and L5 to L1–L3. If there are K families of predicates with the ith family containing n_i predicates, then define an array

$$X = [X_{i,j}]$$

having K rows with the ith row being a permutation of $(1, ..., n_i)$. The set of all such arrays will be denoted $\mathscr{X}$. The array $X \in \mathscr{X}$ applied to the families of predicates $\{\mathscr{P}_i\}$ with $\mathscr{P}_i = \{P_{i,1}, ..., P_{i,n_i}\}$ induces a new family $\{\mathscr{P}_i'\}$ with

$$P'_{i,j} = P_{i,x_{i,j}}.$$

Furthermore, X induces a permutation of the complete description $\mathscr{Q}$ with

$$Q_i = \bigwedge_j P_{ij} \to \bigwedge_j P_{i,x_{i,j}} = Q_{\pi_i}.$$

The set of all such permutations π of $\mathscr{Q}$ induced by some $X \in \mathscr{X}$ will be denoted $\pi_{\mathscr{Q}}$. We can now state

Lemma 2. If C satisfies L1–L4, then m is characterized by $(*)$ with the proviso that the measure F has the invariance property

$$(\forall \pi \in \pi_{\mathscr{Q}}) F(dp_1, ..., dp_q) = F(dp_{\pi_1}, ..., dp_{\pi_q}).$$

Lemma 3. If C satisfies L1–L5, then m is characterized as in Lemma 2 with the additional constraint on F that

$$(\forall r, s)\, n_r = n_s, \qquad X = [X_{i,j}], \qquad X' = [X'_{i,j}],$$

$$X'_{i,j} = \begin{cases} X_{i,j} & \text{if } i \neq r, s \\ X_{r,j} & \text{if } i = s \\ X_{s,j} & \text{if } i = r, \end{cases}$$

π and π' the permutations induced on $\mathscr{Q}$ by X and X' (respectively),

$$F(dp_{\pi_1}, ..., dp_{\pi_q}) = F(dp_{\pi_1'}, ..., dp_{\pi_q'}).$$

The role of L6 is more readily indicated than were the roles of L4 and L5.

Lemma 4. If C satisfies L1–L3 and L6, then m satisfies $(*)$ with the proviso that F is symmetric, that is,

$$(\forall \{\pi_i\} \text{ permutation of } \{1, ..., q\}) F(dp_1, ..., dp_q) = F(dp_{\pi_1}, ..., dp_{\pi_q}).$$

Let $\{0^i\}$ denote the rank ordered $(0^i \geqslant 0^{i+1})$ set of frequencies $\{0_i\}$.

Corollary. If C satisfies L1–L3 and L6, then there is a unique, symmetric measure F such that

$$m\left(\bigwedge_{j=1}^{n} Q_{i_j} a_j\right) = \int \cdots \int \left(\prod_{i=1}^{q} p_i^{0^i}\right) F(dp_1, \ldots, dp_q).$$

Unfortunately, while L6 and L7 are individually attractive they are inconsistent.

Lemma 5. There is no C satisfying L1–L3, L6, and L7.

The set (L1–L5 and L7) is consistent but not categoric. Further axioms are required to define a uniquely "best" confirmation function. Before turning to such axioms, we observe that in the impossibility of substituting L6 for L4 and L5, when L7 is desired, lies an inadequacy of logical probability that cannot be corrected by subsequent axioms. Whatever d.c. we are led to will unavoidably assign degrees of confirmation in a manner that depends on features of $\mathscr{L}$ which may be irrelevant to the meaning of the hypothesis and evidence statements. That is to say, changing the language $\mathscr{L}$ without apparently changing the meaning of H and E will lead to a change in $C(H/E)$.

4. Learning from Experience

Carnap supplemented the characterization of d.c. by postulating properties of C which govern its behavior in certain instances of inductive reasoning. Unfortunately, of the three axioms he proposed, the first is implied by the second, the second does little to narrow the class of acceptable d.c., and the third is merely Johnson's sufficiency postulate (Subsection IVJ2, axiom A2) previously discussed in connection with relative-frequency.

The first axiom concerns the property of instantial relevance, and in a weak form is given by

L8a. $C\left(Q_k a_{n+2}/Q_k a_{n+1} \wedge \left(\bigwedge_{j=1}^{n} Q_{i_j} a_j\right)\right) \geqslant C\left(Q_k a_{n+2}/\bigwedge_{j=1}^{n} Q_{i_j} a_j\right).$

This axiom asserts that confirmation does not decrease when we add confirming instances to the evidence.

Lemma 6.   If C satisfies L1–L3, then C satisfies L8a.

A strict form of instantial relevance would be

L8b. $C\left(Q_k a_{n+2}/Q_k a_{n+1} \wedge \left(\bigwedge_{j=1}^{n} Q_{i_j} a_j\right)\right) > C\left(Q_k a_{n+2}/\bigwedge_{j=1}^{n} Q_{i_j} a_j\right).$

Lemma 7.   If C satisfies L1–L3, then C satisfies L8b if and only if the measure F in the characterization ($*$) of m has nondegenerate marginal distributions.

Hence, even strict instantial relevance does little to restrict the set of possible confirmation functions. The reader should also consult [22, pp. 228–251] for related analyses of which we were unaware.

A second axiom concerns the asymptotic behavior of C as the number n of individuals $\{a_i\}$ increases. Let $0_k(n)$ be the number of occurrences of Q_k in $(Q_{i_1}, ..., Q_{i_n})$.

L9. $(\forall\{Q_{i_j}\}) \lim_{n\to\infty} \left[C\left(Q_k a_1/\bigwedge_{j=2}^{n} Q_{i_j} a_j\right) - \frac{0_k(n)}{n} \right] = 0.$

The significance of L9 is immediate from

Lemma 8.   If C satisfies L1–L3, then C satisfies L9 if and only if the measure F in ($*$) has as support the set

$$R_q = \left\{(p_1, ..., p_q) : p_i \geqslant 0, \sum_{i=1}^{q} p_i = 1\right\}.$$

We note that L9 does not considerably narrow the class of possible confirmation functions. Furthermore L6 is now seen to imply L8b, and therefore the pair of axioms is no more useful than is L9 alone.

The final axiom is the Johnson sufficiency postulate [14].

L10. $(\exists g)\, C\left(Q_k a_{n+1}/\bigwedge_{j=1}^{n} Q_{i_j} a_j\right) = g(0_k, n).$

The exact distribution of occurrences of predicates other than Q_k is irrelevant with respect to the confirmation of $Q_k a_{n+1}$. Carnap asserts that when there is only one family $\mathscr{P}_1$ of n_1 predicates in $\mathscr{L}$, then the d.c. satisfying L1–L4 and L10 can be characterized as follows:

$$(\exists 0 < \lambda < \infty)\, C_\lambda\left(Q_k a_{n+1}/\bigwedge_{j=1}^{n} Q_{i_j} a_j\right) = \frac{0_k + \lambda/n_1}{n + \lambda}.$$

This one-parameter family of d.c. $\{C_\lambda\}$ comprises what Carnap grandly called the continuum of inductive methods [15].

To derive the measure m_λ corresponding to C_λ, note that if T is any tautology, then

$$m_\lambda \left(\bigwedge_{j=1}^{n} Q_{i_j} a_j \right) = C_\lambda \left(\bigwedge^{n} Q_{i_j} a_j / T \right) = C_\lambda (Q_{i_1} a_1 / T) \prod_{r=1}^{n-1} C \left(Q_{i_{r+1}} a_{r+1} / \bigwedge_{j=1}^{r} Q_{i_j} a_j \right).$$

From the definition of C_λ we have that

$$C_\lambda (Q_{i_1} a_1 / T) = \frac{1}{q},$$

$$C_\lambda \left(Q_{i_{r+1}} a_{r+1} / \bigwedge^{r} Q_{i_j} a_j \right) = \frac{0_{i_{r+1}}(r) + \lambda/q}{r + \lambda},$$

where $0_{i_{r+1}}(r)$ is the number of individuals in $(a_1, ..., a_r)$ that are described by $Q_{i_{r+1}}$. It is now immediate that

$$m_\lambda \left(\bigwedge_{j}^{n} Q_{i_j} a_j \right) = \frac{1}{q} \left(\prod_{r=1}^{n-1} [r + \lambda]^{-1} \right) \prod_{r=1}^{n-1} \left[0_{i_{r+1}}(r) + \frac{\lambda}{q} \right].$$

The properties of m_λ become more apparent after rewriting. Let $\{0^i\}$ be the rank ordered $(0^i \geqslant 0^{i+1})$ set of frequencies of occurrence $\{0_i\}$. With the understanding that $\prod_{i=1}^{0} f(i) = 1$, we find that

$$m_\lambda \left(\bigwedge_{j}^{n} Q_{i_j} a_j \right) = \frac{1}{q} \left(\prod_{r=1}^{n-1} [r + \lambda]^{-1} \right) \prod_{j=1}^{q} \left(\prod_{i=1}^{0^j} \left[i + \frac{\lambda}{q} \right] \right).$$

5. Selection of a Unique Confirmation Function

If we have only a single family of predicates, and we follow Carnap, then we can conclude with a representation for confirmation functions that involves one free parameter $\lambda \in (0, \infty)$. Selection of a unique C_λ confirmation function requires us to choose λ. Carnap [15, pp. 56–79] has indicated several arguments that might form the basis for a selection of λ, although these arguments do not all lead to the same choice of λ. One interesting argument is that λ reflects our *a priori* anticipations as to the balance between the occurrences of individuals satisfying each of the n_1 predicates in the single family $\mathscr{P}_1$. Small values of λ correspond to anticipating that the predicates $\{P_{1i}\}$ do not occur equally often in descriptions of $\{a_j\}$, whereas large values of λ correspond to anticipating a balance in the frequencies of occurrence of the $\{P_{1i}\}$ in descriptions of the individuals $\{a_j\}$. Such a vague argument, however, leaves much to be desired in an *a priori* theory of rational induction and rational decision-making.

In his earlier work Carnap [4, pp. 562–577] had proposed, albeit with little enthusiasm, a confirmation function C^* as the most appropriate choice; C^* is defined through m^* by

$$m^* \left(\bigwedge_{j=1}^{n} Q_{i_j} a_j \right) = \frac{(q-1)! \prod_{j=1}^{q} (0_j!)}{(n+q-1)!} .$$

C^* is easily seen to satisfy L1–L6. We may interpret m^* as arising from a desire to assign equal weight to all state descriptions having the same set of ranked frequencies $\{0^i\}$ and equal weight to all statements (called structure descriptions) of the form

$$\{0^i\} = \{m_i\}.$$

The measures m^* and m_λ agree in assigning equal weight to all statements having the same structure descriptions $\{0^i\}$ but disagree in their assignments of weights to different structure descriptions.

The grounds for preferring one confirmation function to another as the explication of correct or rational inductive inference or as the basis for rational decision-making include the intuitive or *a priori* acceptability of the axioms used to characterize the "best" confirmation function as well as our willingness to forgo those properties of a desirable confirmation function that the "best" function does not possess. As we have seen in the discussion of L6 and L7 there does not exist a function having all of what might be thought to be the desirable properties of confirmation. With respect to the choice of axioms for inductive logic, Carnap[†] has commented:

(a)   The reasons are based upon our intuitive judgments concerning inductive validity, i.e., concerning inductive rationality of practical decisions (e.g., about bets).
Therefore:
(b)   It is impossible to give a purely deductive justification of induction.
(c)   The reasons are *a priori* [10, p. 978].

The interested reader can find arguments for the acceptability of the axioms we have presented in the referenced writings of Carnap and Kemeny.

[†] P. A. Schilpp, ed., *The Philosophy of Rudolf Carnap*, p. 978. La Salle, Illinois: The Open Court Publishing Co., 1963.

VIIE. Logical Probability and Relative-Frequency

Carnap has suggested that while logical probability attaches only to inferences, it can also serve as a rational estimate of empirical probability, especially in the relative-frequency interpretation. A language $\mathscr{L}$ to describe the model of unlinked, indefinitely repeated experiments would contain individuals $\{a_i\}$, where a_i is the outcome of the ith experiment, and a single family $\mathscr{P}$ of predicates $\{Q_i\}$ that completely describe the possible outcomes of the experiment. An evidence statement might be of the form $\bigwedge_{j=1}^{n} Q_{i_j} a_j$, and we might be interested in assessing the probability of the hypothesis $Q_k a_{n+1}$. The hypotheses of unlinkedness and indefinite repeatability seem to justify L2 and L3 for C. It is then immediate from Lemma 8 that if C satisfies L1–L3 and the measure F in $(*)$ has R_q as support, then

$$\lim_{n \to \infty} \left[C \left(Q_k a_{n+1} \bigg/ \bigwedge_{j=1}^{n} Q_{i_j} a_j \right) - \frac{0_k(n)}{n} \right] = 0.$$

Hence, if the frequency $0_k(n)/n$ of outcomes Q_k converges to ρ_k, then so will C; that is, logical probability will approximate relative-frequency-based empirical probability when the latter is "appropriate."

If we do not wish to assume that F has R_q as support and/or are wisely not content with asymptotic arguments, then Carnap has proposed a less direct link between logical and relative-frequency-based probability. We first define a logical probability version of an estimator.

Definition. If a function f of the states of the world takes on the possible values $\{f_i\}$ and H_i is the hypothesis that $f = f_i$, then the estimate of f relative to evidence E based on a confirmation function C is

$$e(f, E, C) = \sum_i f_i C(H_i/E).$$

These estimates are similar to expected values, and their merits are argued by Carnap [4, Chapter IX]. If, for example, we have a set of statements $L_1, ..., L_r$ and wish to estimate the proportion f of true statements in the set, given evidence E and a confirmation function C satisfying L1, then letting H_m be the hypothesis that exactly m of $\{L_i\}$ are true we have

$$e(f, E, C) = \sum_{m=1}^{r} \frac{m}{r} C(H_m/E).$$

Carnap [4, p. 543] proves the following.

Theorem 3. Under the preceding definition of terms,

$$\sum_{m=1}^{r} mC(H_m/E) = \sum_{i=1}^{r} C(L_i/E).$$

The estimate of the proportion of true statements in a set can be determined directly from the degrees of confirmation of the statements.

We can now relate degree of confirmation to the estimation of relative-frequency-based empirical probability. Let there be r trials; let L_i be the statement of success on trial i; and let E be an evidence statement implying that the outcomes of the trials are uninfluenced by their order, that is, $(\forall i)\ C(L_i/E) = C(L_1/E)$. It is now apparent from Theorem 3 that $C(L_1/E)$ is the estimate of the relative-frequency of success in the r trials. Furthermore, since this conclusion is independent of r, we see that for suitable E, $C(L_1/E)$ is also an estimate of the limit of relative-frequency as r goes to infinity. Note that this estimate does not require the assumption that the actual relative-frequency of success has a limit. Thus $C(L_1/E)$ is an estimate of relative-frequency-based empirical probability in the same sense that $\sum_{m=1}^{r} mC(H_m/E)$ is an estimate of the number of true $\{L_i\}$.

Clearly, since any C satisfying $L1$ can serve to calculate an estimate of relative-frequency-based probability, the "estimate" is in nowise bound to be empirically correct. Theorem 3 is only a mathematical tautology concerning a suggestively labeled quantity and cannot, despite appearances, be expected to yield empirically significant conclusions.

VIIF. Applications of C^* and C_λ

As a first illustration of the use of quantitative logical probability consider an experiment such as repeated die tossing in which there is one family $\{Q_i\}$ of predicates to describe the outcome of each of the repeated trials $\{a_i\}$. Let a success Sa_i at trial i be defined by

$$Sa_i = \bigvee_{k=1}^{w} Q_{j_k} a_i;$$

a success occurs if any of w outcomes described by $Q_{j_1}, ..., Q_{j_w}$ occurs. Given evidence E of the form

$$E = \bigwedge_{j=1}^{n} Q_{i_j} a_j,$$

we may be interested in an hypothesis H of the form Sa_{n+1}. If there were s successes in the first n trials, then recourse to C^* yields,

$$C^*(H/E) = (s + w)/(n + q).$$

The form of this conditional probability for a future success is identical to that of the Bayes estimator $\hat{p}$ of the probability p of success calculated using a quadratic loss function ($\hat{p} - p)^2$ and a Beta prior density for p; that is,

$$\frac{(q - 1)!}{(w - 1)!(q - w - 1)!} p^{w-1}(1 - p)^{q-w-1} \qquad \text{for} \quad p([0, 1]).$$

From the representation for m_λ given at the end of Subsection VIID4, we can easily calculate that

$$C_\lambda(H/E) = (s + w\lambda/q)/(n + \lambda).$$

Clearly, for this problem choosing $\lambda = q$ would produce agreement between C^* and C_λ as to the degree of confirmation of the hypothesis of a success at a future trial given the record of past outcomes. Furthermore, C_λ can also be viewed as the Bayes estimator $\hat{p}$ of the probability p of success calculated using a quadratic loss and Beta prior density for p. Hence both C^* and C_λ are not unreasonable confirmation functions in that they are statistically admissible for some problem.

As an additional illustration of the use of C^* and C_λ we consider the following communications problem. A source or transmitter, about which we have little prior knowledge, generates messages that are binary sequences of length N. These messages are communicated through a poorly understood channel and received as possibly different binary sequences of length N. We collect data in the form of M pairs of transmitted and received sequences; the pair may be thought of as a binary sequence of length $2N$ with the initial segment the transmitted sequence. We are now informed that a given $(M + 1)$th sequence has been received and wish to infer what has been transmitted. For example we may wish to know the degree of confirmation for the hypothesis that more 0's were transmitted than 1's. A formulation of this problem in the terms of logical probability is as follows.

There are $M + 1$ individuals $\{a_i\}$ representing the pairs of transmitted and received binary sequences of length N. There are 2^{2N} predicates $\{Q_i\}$, each one specifying one of the possible binary sequences of length $2N$. Let $\{Q_{j_k}\}$ denote the 2^N predicates describing each possible sequence of length $2N$ which terminates in the observed received sequence and $\{Q_{i_j}\}$

describe the M known pairs of transmitted–received sequences. Then the range $\mathscr{R}(E)$ of the evidence or data is

$$\mathscr{R}(E) = \left\{ S : (\exists Q_{j_k} \in \{Q_{j_r}\})\ S = Q_{j_k} a_{M+1} \wedge \left(\bigwedge_{j=1}^{M} Q_{i_j} a_j \right) \right\}.$$

The range $\mathscr{R}(H \wedge E)$ of the conjunction of hypothesis H [more 0's sent than 1's in the $(M+1)$th message] and evidence E is the subset of $\mathscr{R}(E)$ for which Q_{j_k} described only those sequences where there are more 0's than 1's in the first N symbols and the observed $(M+1)$th received sequence for the remaining N symbols. If n_1 is the number of the M observed transmitted–received sequences for which the first N symbols contained more 0's than 1's and the last N symbols agreed with the last N symbols of the $(M+1)$th sequence (whose initial segment we are inferring), and n_2 is the number of observed sequences agreeing in the last N symbols with the $(M+1)$th sequence, then use of m^* yields

$$C^*(H/E) = \begin{cases} \dfrac{n_1 + 2^{N-1}}{n_2 + 2^N} & \text{if } N \text{ is odd,} \\[2ex] \dfrac{n_1 + \frac{1}{2}\left(2^N - \binom{N}{N/2}\right)}{n_2 + 2^N} & \text{if } N \text{ is even.} \end{cases}$$

To calculate $C_\lambda(H/E)$ through m_λ as represented at the end of Subsection VIID4 we need to introduce the following.

n_3 number of distinct Q_j in E such that Q_j describes a sequence with more 0's sent than 1's and a received sequence that agrees with that of a_{M+1}.

n_4 number of distinct Q_j in E such that Q_j describes a sequence in which the received sequence agrees with that of a_{M+1}.

Recourse to m_λ yields

$$C_\lambda(H/E) = \frac{n_1 + \lambda 2^{-2N} n_3}{n_2 + \lambda 2^{-2N} n_4}.$$

Hence we have found logical probabilities for the $(M+1)$th transmitted sequence to contain more 0's than 1's given the received sequence and M transmitted–received pairs of sequences. Both C^* and C_λ yield probabilities that differ from a straight relative-frequency assessment of n_1/n_2.

VIIG. Critique of Logical Probability

1. Roles for Logical Probability

Possible roles for logical probability include

(1) formalization of inductive reasoning via a measure of inferential support,

(2) source of rational estimates of empirical probability,

(3) explication of classical probability,

(4) basis for rational decision-making.

Too little is known about the modal and comparative concepts of logical probability to discuss them here. Extensive discussions as to whether present concepts of quantitative logical probability can, or do, fulfill any of these roles are available in Hempel [7], Kyburg [16], and Schilpp [10], and are not reproduced here. While we will indicate a few objections to the way present concepts of logical probability fulfill the above-mentioned roles, it is more difficult to determine whether these objections must bear against any concept of logical probability.

A serious objection to Carnap's concept of logical probability is that it is not applicable to the full range of uses of inductive reasoning. The limited languages that are treated by Carnap are unequal to the description of most of the usual scientific observations such as those involving numerical measurements. Hence we cannot apply Carnap's logical probability to discuss the degree to which laboratory observations support an hypothesis or theory. Furthermore, in Carnap's formulation all "laws" (propositions involving universal quantifiers) always have a zero degree of confirmation given any finite number of supporting instances. Albeit, universal statements being of no importance in the formulation of real problems, we are not inclined to take this latter defect of the Carnapian concept of logical probability seriously.

Reformulations of logical probability by Scott and Krauss and by Hintikka promise to avoid the aforementioned difficulties. Scott and Krauss [17] and Krauss [18] avail themselves of model theory in an attempt to define logical probability for richer and more realistic languages. Hintikka [19] has developed a version of logical probability wherein universal statements can have positive degrees of confirmation.

While we feel that logical probability may serve to explicate an objective version of classical probability, through an algorithmic assessment of the support provided by statements of prior knowledge for the various alternative outcomes, we are less inclined to agree that logical probability is the proper basis for estimating empirical (say relative-frequency-based) probability. There is a significant informal and subjective component in the selection of good estimates that may be uneliminatable; an experimenter can rarely write out in some simple language all he knows about an experiment.

There may be a role for logical probability in rational decision-making, although the form it will take is as yet unclear. The Anscombe–Aumann development of subjective probability (Section VIIIC) suggests one way of linking logical probability with decision-making. However, this link has yet to be detailed and the resulting relationship justified.

2. Formulation of Logical Probability

Questions of formulation concern the axiomatic structure of logical probability and the manner in which a choice is made, consistent with the axioms, of a logical probability relation to resolve a specific problem of inductive inference. Our discussion of modal logical probability completely ignored the problem of selecting a specific modal quantifier $\mathscr{P}$ and thus represents but a fragment of a useful theory. Koopman supplied us not only with a weak axiomatization of comparative logical probability but also with the directive that the specific choice be made subjectively. However, in contrast to the development of subjective probability in Chapter VIII, where the choice of subjective measure is guided by the desire for good decisions, Koopman in no way guides our subjective selection of order relations. At best Koopman has motivated and supplied a starting point for a useful theory of comparative logical probability.

Carnap and others attempted axiomatic specifications of quantitative logical probability or degree of confirmation (d.c.). Carnap's original objective of a categoric axiomatization to uncover the unique logically necessary d.c. would have resolved the problem of a specific choice of relation. However, the goal of a logically necessary d.c. seems to have been abandoned in the face of criticisms that were leveled at all of the candidates. Furthermore, results such as Lemma 5 suggest that there cannot exist a "best" d.c. function. The many desirable properties of rational inductive inference procedures, as reflected in the axioms for logical probability, are inconsistent. Failing a "best" choice of d.c., we need to know more about the bases for choosing between those d.c. functions satisfying the agreed upon axioms so as to resolve specific problems of inductive inference properly. There has been some discussion of this question, particularly as it affects C_λ [15, pp. 56–79]. However, we expect that it will prove to be very difficult to illuminate this subject.

Turning to the axiomatic formulations themselves, we have previously noted the existence of systems of logical probability intended to overcome certain problems with Carnap's construction [17, 18]. A generic difficulty with these various formulations is the dependence of the resulting d.c. on the choice of language in which the evidence and hypothesis are expressed. Syntactical considerations intrude on questions that seem to be purely semantical. (This difficulty is analogous to the irksome dependence of complexity evaluations on the arbitrary choice of AUTM.) Possibly though this conflict with Salmon's seemingly reasonable criterion of linguistic invariance (Subsection IVJ2) is inevitable and its direct confrontation a virtue of the approaches to logical probability.

3. Justifying Logical Probability

A widely espoused criticism of logical probability is, in the words of Black [20],

> The most difficult question that any "logical" theory has to answer is how *a priori* truths can be expected to have any bearing upon the practical problem of anticipating the unknown on the basis of nondemonstrative reasons.

Carnap responded to this objection by pointing out that the degree of confirmation of a statement by an evidence statement E is not *a priori* in that it contains the synthetic content of our asserting E to be true. We might also note the relevance, to a defense of logical probability, of the difficulties encountered in selecting a unique d.c. While Carnap desires a definition of a d.c. that can be stated in purely logical terms without reference to contingent matters of fact, the choice of a best d.c. seems to involve extralogical, intuitive considerations. These considerations are at least in part to be distilled from our awareness of the purposes of a theory of probability as well as our extensive experience with inductive reasoning and are not *a priori*; *vide* the arguments for choosing λ in C_λ [15, pp. 56–79]. Of course, this defense is also an attack against the conception that the best d.c. could be selected on logical or analytic grounds alone. Perhaps part of the difficulty here stems from the dubious assumption of a dichotomy between the analytic and the synthetic [21].

It is essential that the relations between logical probability, empirical probability, and practical concerns be clarified and strengthened, if possible. We noted in Section VIIE that logical probability is compatible with relative-frequency-based empirical probability. Albeit our studies in Chapter IV and elsewhere lead us to take little comfort from this. More significant has been Carnap's interest in pragmatically justifying logical probability as leading to a fair betting quotient, or better yet, as appropriate for calculating the expected utilities of decisions. However, the justification of logical probability through a role in rational decision-making does not seem to have been fully argued, the several invocations of such a justification notwithstanding. Strict coherence, when possible, and it is not always possible, may be necessary, but it is hardly a sufficient guide to decision-making. Overall, Carnap's conception of logical probability seems most nearly to be a clarification and refinement of classical probability.

The attempt to base a theory of probability on its role in decision-making is at the root of the theories of subjective probability. As we see in Chapter VIII, subjective theories typically admit the use of individual judgment, as did Koopman, and do not yield the uniquely "correct" probability distribution that was Carnap's original goal.

VII. Appendix: Proofs of Results

Lemma 1. If C satisfies L1–L3, then there is a unique measure F such that

$$R_q = \left\{ (p_1, ..., p_q) : p_i \geqslant 0, \sum_{i=1}^{q} p_i = 1 \right\},$$

$$F(R_q) = 1,$$

$$m \left(\bigwedge_{j=1}^{n} Q_{i_j} a_j \right) = \int \cdots \int \left(\prod_{j=1}^{n} p_{i_j} \right) F(dp_1, ..., dp_q),$$

and q is the number of complete descriptions.

Proof. By L1, C can be represented by a measure m. By L2 and L3 true for all n, m is an exchangeable measure. The representation of m is that stated for exchangeable measures by de Finetti and Hewitt and Savage [13].

Lemma 2. If C satisfies L1–L4, then m is characterized by $(*)$ with the proviso that the measure F has the invariance property

$$(\forall \pi \in \pi_2) F(dp_1, ..., dp_q) = F(dp_{\pi_1}, ..., dp_{\pi_q}).$$

Proof.   From L4

$$(\forall \pi \in \pi_2) \, m \left(\bigwedge_{j=1}^{n} Q_{i_j} a_j \right) = m \left(\bigwedge_{j=1}^{n} Q_{\pi_{i_j}} a_j \right).$$

Hence by $(*)$

$$\int \cdots \int \left(\prod_{i=1}^{q} p_i^{0_i}\right) F(dp_1, ..., dp_q) = \int \cdots \int \left(\prod_{i=1}^{q} p_{\pi_i}^{0_{\pi_i}}\right) F(dp_1, ..., dp_q).$$

Note that each permutation $\pi \in \pi_{\mathscr{Q}}$ has an inverse $\pi^{-1} \in \pi_{\mathscr{Q}}$. Changing variables in the right-hand side of the preceding equation yields

$$\int \cdots \int \left(\prod_{i=1}^{q} p_i^{0_i}\right) F(dp_1, ..., dp_q) = \int \cdots \int \left(\prod_{i=1}^{q} p_i^{0_i}\right) F(dp_{\pi_1^{-1}}, ..., dp_{\pi_q^{-1}}).$$

Since the measure F is unique, the lemma follows with π^{-1} in place of π. ∎

Lemma 3. If C satisfies L1–L5, then m is characterized as in Lemma 2 with the additional constraint on F that

$$(\forall r, s) \, n_r = n_s, \qquad X = [X_{i,j}], \qquad X' = [X'_{i,j}],$$

$$X'_{i,j} = \begin{cases} X_{i,j} & \text{if } i \neq r, s \\ X_{r,j} & \text{if } i = s \\ X_{s,j} & \text{if } i = r, \end{cases}$$

π and π' the permutations induced on $\mathscr{Q}$ by X and X' (respectively),

$$F(dp_{\pi_1}, ..., dp_{\pi_q}) = F(dp_{\pi_1'}, ..., dp_{\pi_q'}).$$

Proof. Parallels that of Lemma 2 and is omitted. ∎

Lemma 4. If C satisfies L1–L3 and L6, then m satisfies $(*)$ with the proviso that F is symmetric, that is,

$$(\forall \{\pi_i\} \text{ permutation of } \{1, ..., q\}) \, F(dp_1, ..., dp_q) = F(dp_{\pi_1}, ..., dp_{\pi_q}).$$

Proof. This lemma is a corollary to Lemma 2 restricted to the case of only one family of predicates. ∎

Lemma 5. There is no C satisfying L1–L3, L6, and L7.

Proof. The assumption that we can adjoin arbitrarily many families of predicates to $\mathscr{L}$ and still leave $m(\bigwedge Q_{i_j} a_j)$ invariant corresponds, through Lemma 4, to assuming the existence of an infinite family of consistent, symmetric measures $\{F_q(dp_1, ..., dp_q)\}$. Hence from the de Finetti and Hewitt and Savage [13] characterization of symmetric (exchangeable) families of measures we have the following.

There exist a probability measure H and a conditional probability measure G such that

$$(\forall q)(\forall A) F_q(A) = \int H(d\alpha) \int \cdots \int_A \prod_{i=1}^{q} G(dp_i \mid \alpha).$$

However, we assert that this characterization of F_q is incompatible with the requirement

$$(\forall q) F_q(R_q) = 1.$$

To verify the asserted incompatibility, consider the following mutually exclusive and exhaustive possibilities:

(1) $(\exists A)(H(A) > 0)(\forall \alpha \in A) \int_{1/q}^{1} G(dp \mid \alpha) > 0;$

(2) $(\exists A)(H(A) > 0)(\forall \alpha \in A) \int_{0}^{1/q} G(dp \mid \alpha) > 0;$

(3) $(\exists A)(H(A) = 1)(\forall \alpha \in A) G\left(p = \dfrac{1}{q} \,\Big|\, \alpha\right) = 1.$

Under (1),

$$F_q\left(\left\{(p_1, ..., p_q) : \sum_{i=1}^{q} p_i > 1\right\}\right) > \int_A H(d\alpha) \left\{\int_{1/q}^{\infty} G(dp \mid \alpha)\right\}^q > 0.$$

Similarly, under (2),

$$F_q\left(\left\{(p_1, ..., p_q) : \sum_{i=1}^{q} p_i < 1\right\}\right) > 0.$$

Both of these conclusions conflict with $F_q(R_q) = 1$, leaving only (3). However, (3) is not invariant with respect to the number q of complete descriptions and therefore violates L7. ∎

Lemma 6. If C satisfies L1–L3, then C satisfies L8a.

Proof. From the characterization $(*)$ of m and introducing the notation

$$X = \prod_{i \neq k} p_i^{0i}, \qquad Y = p_k,$$

$$EXY^n = \int \cdots \int \left(\prod_{i \neq k} p_i^{0i}\right) p_k^{n} F(dp_1, ..., dp_q),$$

we have that

$$C\left(Q_k a_{n+2}/Q_k a_{n+1} \wedge \left(\bigwedge_{j=1}^{n} Q_{i_j} a_j\right)\right) = EXY^{n+2}/EXY^{n+1},$$

$$C\left(Q_k a_{n+2}/\bigwedge_{j=1}^{n} Q_{i_j} a_j\right) = EXY^{n+1}/EXY^n.$$

Note that the random variables X and Y are nonnegative. Hence, by Schwarz's inequality,

$$E^2 XY^{n+1} = E^2(\sqrt{XY^{n+2}})(\sqrt{XY^n}) \leqslant E(XY^{n+2}) E(XY^n).$$

It is immediate that

$$C\left(Q_k a_{n+2}/Q_k a_{n+1} \wedge \left(\bigwedge_{j=1}^{n} Q_{i_j} a_j\right)\right) \geqslant C\left(Q_k a_{n+2}/\bigwedge_{j=1}^{n} Q_{i_j} a_j\right). \quad \blacksquare$$

Lemma 7. If C satisfies L1–L3, then C satisfies L8b if and only if the measure F in the characterization (∗) of m has nondegenerate marginal distributions.

Proof. Examination of the proof of Lemma 6 shows that equality between confirmations occurs only when there is equality in the Schwarz inequality

$$E^2(\sqrt{XY^{n+2}})(\sqrt{XY^n}) \leqslant E(XY^{n+2}) E(XY^n).$$

As is well known, there is equality in the Schwarz inequality only if there is linear dependence; that is,

$$(\exists \alpha) \sqrt{XY^{n+2}} = \alpha\sqrt{XY^n} \qquad \text{(a.e. } F\text{)}.$$

Hence the condition for equality between the confirmations is that

$$(\forall k) \, p_k \text{ is a constant} \qquad \text{(a.e. } F\text{)}.$$

The lemma follows immediately. $\quad \blacksquare$

Lemma 8. If C satisfies L1–L3, then C satisfies L9 if and only if the measure F in (*) has as support the set

$$R_q = \left\{(p_1, ..., p_q) : p_i \geqslant 0, \sum_{i=1}^{q} p_i = 1\right\}.$$

Proof. We first show that if F as R_q as support and m satisfies L1–L3, then m satisfies L9. Define

$$f(\mathbf{p}) = \sum_{i=1}^{q} p_i \log p_i ,$$

where $p_i = 0_i/n$ and $\mathbf{p}$ denotes the vector $(p_1 , ..., p_q)$. It follows from the continuity and concavity of $\log x$ that f is continuous and concave in the interior of R_q. It is easily shown that f has a unique maximum in R_q at $\mathbf{p} = \boldsymbol{\rho}$, where $\boldsymbol{\rho}$ denotes the vector $(\rho_1 , ..., \rho_q)$. Furthermore, for all $\mathbf{q}$ and $0 \leqslant \lambda \leqslant 1$, $f(\lambda\mathbf{q} + (1 - \lambda)\boldsymbol{\rho})$ is strictly decreasing in λ. To verify these statements, note that

$$\frac{\partial f(\lambda\mathbf{q} + (1 - \lambda)\boldsymbol{\rho})}{\partial \lambda} = \sum_{i=1}^{q} \frac{\rho_i(q_i - \rho_i)}{\lambda q_i + (1 - \lambda)\rho_i} ,$$

and write

$$\rho_i = -\lambda(q_i - \rho_i) + (1 - \lambda)\rho_i + \lambda q_i$$

to find

$$\frac{\partial f}{\partial \lambda} = \sum_{i=1}^{q} (q_i - \rho_i) - \lambda \sum_{i=1}^{q} \frac{(q_i - \rho_i)^2}{\lambda q_i + (1 - \lambda)\rho_i} .$$

Noting that as $\mathbf{q}$ and $\boldsymbol{\rho}$ are both in R_q,

$$\sum (q_i - \rho_i) = 0,$$

we see that $\partial f/\partial \lambda$ is strictly negative unless $\mathbf{q} = \boldsymbol{\rho}$.

An inequality that will be useful to us is derived as follows. Note that

$$(\forall \mathbf{q}, \boldsymbol{\rho} \in R_q) \ \lambda q_i + (1 - \lambda)\rho_i \leqslant 1,$$

to conclude that

$$\frac{\partial f}{\partial \lambda} \leqslant -\lambda \sum_{i=1}^{c} (q_i - \rho_i)^2.$$

Hence,

$$f(\mathbf{q}) = f(\boldsymbol{\rho}) + \int_0^1 \frac{\partial f}{\partial \lambda} \, d\lambda \leqslant f(\boldsymbol{\rho}) - 1/2 \sum_{i=1}^{q} (q_i - \rho_i)^2.$$

Define the open ball $B(\boldsymbol{\rho}, \epsilon)$ in R_q with center $\boldsymbol{\rho}$ and radius ϵ by

$$B(\boldsymbol{\rho}, \epsilon) = \left\{ (p_1 , ..., p_q) : p_i \geqslant 0, \sum_{i=1}^{q} p_i = 1, \sum_{i=1}^{q} (p_i - \rho_i)^2 < \epsilon^2 \right\}.$$

From the preceding inequality we conclude that

$$(\forall \rho, \epsilon) \sup_{p \notin B(\rho, \epsilon)} f(\mathbf{p}) \leqslant f(\rho) - \epsilon^2/2.$$

Define

$$R^\delta = \{(p_1, ..., p_q) : (\forall_i)\, p_i \geqslant \delta\} \cap R_q.$$

From the continuity of f on R^δ and the compactness of R^δ we have

$$(\forall \epsilon > \delta > 0)(\exists \gamma_0 > 0)(\forall \rho \in R_q)(\forall \mathbf{p} \in B(\rho, \gamma_0) \cap R^\delta)\, f(\mathbf{p}) > f(\rho) - \epsilon^2/4.$$

Furthermore, from the hypothesis that F has R_q as support we can choose γ_0 such that

$$(\forall \epsilon > \delta > 0)(\exists c_{\gamma_0, \delta})(\forall \rho)\, F\big(B(\rho, \gamma_0) \cap R^\delta\big) \geqslant c_{\gamma_0, \delta} > 0.$$

Noting that

$$\prod_{i=1}^{q} p_i^{0_i} = \left[\prod_{i=1}^{q} p_i^{\rho_i}\right]^n = e^{nf(\mathbf{p})},$$

we see that for $0 < \gamma \leqslant \min(\epsilon, \gamma_0)$,

$$\int \cdots \int_{B(\rho, \epsilon)} e^{nf(\mathbf{p})} F(dp_1, ..., dp_q) \geqslant \int \cdots \int_{B(\rho, \gamma)} e^{nf} F(dp_1, ..., dp_q)$$

$$\geqslant c_{\gamma, \delta} \exp(n[\, f(\rho) - \epsilon^2/4]),$$

$$\int \cdots \int_{\overline{B(\rho, \epsilon)}} e^{nf(\mathbf{p})} F(dp_1, ..., dp_q) \leqslant \exp(n[\, f(\rho) - \epsilon^2/2]) F(\bar{B}) \leqslant \exp(n[\, f(\rho) - \epsilon^2/2]).$$

Hence,

$$(\forall \{\rho_n\}) \lim_{n \to \infty} \frac{\int \cdots \int_{\overline{B(\rho_n, \epsilon)}} \exp[nf_n(\mathbf{p})]\, F(dp_1, ..., dp_q)}{\int \cdots \int_{B(\rho_n, \epsilon)} \exp[nf_n(\mathbf{p})]\, F(dp_1, ..., dp_q)} \leqslant \lim_{n \to \infty} \frac{\exp(-n\epsilon^2/4)}{c_{\gamma_0, \delta}} = 0,$$

where

$$\rho_n = \big(\rho_1(n), ..., \rho_q(n)\big), \qquad f_n(\mathbf{p}) = \sum_{i=1}^{q} \rho_i(n) \log p_i.$$

Applying this result to the d.c. C we find that

$$(\forall \epsilon > 0)(\forall \{\rho_n\}) \lim_{n \to \infty} \left[C\left(Q_k a_1 \Big/ \bigwedge_{j=2}^{n} Q_{i_j} a_j\right) \right.$$

$$\left. - \frac{\int \cdots \int_{B(\rho_n, \epsilon)} p_k \exp(nf_n)\, F(dp_1, ..., dp_q)}{\int \cdots \int_{B(\rho_n, \epsilon)} \exp(nf_n)\, F(dp_1, ..., dp_q)} \right] = 0.$$

It follows upon taking ϵ arbitrarily small that

$$\lim_{n \to \infty} \left[C \left(Q_k a_1 \Big/ \bigwedge_{j=2}^{n} Q_{i_j} a_j \right) - \rho_k(n) \right] = 0,$$

and the sufficiency of the hypothesis is now apparent.

To verify the necessity that F have R_q as support we need only assume to the contrary that

$$(\exists \rho, \epsilon) \, F(B(\rho, \epsilon)) = 0.$$

It can now be easily shown by examining a sequence $\{Q_{i_j}\}$ such that $\rho_i(n) \to \rho_i$ that

$$C - \rho_k(n) \nrightarrow 0.$$

We omit the details. ∎

The proof of the preceding lemma could also have been recast in probabilistic terms involving the convergence of $E(p_k \mid \{O_i(n)\})$.

References

1. J. M. Keynes, *A Treatise on Probability*. New York: Harper, 1962.
2. H. Jeffreys, *Theory of Probability*. London and New York: Oxford Univ. Press (Clarendon), 1948.
3. B. Koopman, The Bases of Probability, *Bull. Amer. Math. Soc.* **46**, 763–774, 1940.
4. R. Carnap, *Logical Foundations of Probability*, 2nd ed. Chicago, Illinois: Univ. of Chicago Press, 1962.
5. G. H. von Wright, *An Essay in Modal Logic*. Amsterdam: North-Holland Publ., 1951.
6. G. E. Hughes and M. J. Cresswell, *An Introduction to Modal Logic*. London: Methuen, 1968.
7. C. G. Hempel, *Aspects of Scientific Explanation*. Glencoe, Illinois: Free Press, 1965.
8. P. Martin-Löf, The Definition of Random Sequences, *Information and Control* **9**, 619, 1966.
9. B. Koopman, Intuitive Probability and Sequences, *Ann. of Math.* **42**, 185, 1941.
10. P. A. Schilpp, ed., *The Philosophy of Rudolf Carnap*, pp. 969–979. LaSalle, Illinois: Open Court, 1963.
11. P. C. Fishburn, *Utility Theory for Decision Making*. New York: Wiley, 1970.
12. J. G. Kemeny, Fair Bets and Degree of Confirmation, *J. Symbolic Logic* **20**, 263–273, 1955.
13. E. Hewitt and L. J. Savage, Symmetric Measures on Cartesian Products, *Trans. Amer. Math. Soc.* **80**, 484, 489, 1955.
14. I. J. Good, *The Estimation of Probabilities*, p. 26. Cambridge, Massachusetts: MIT Press, 1965.

15. R. Carnap, *The Continuum of Inductive Methods*. Chicago, Illinois: Univ. of Chicago Press, 1952.
16. H. E. Kyburg, Jr., I. Recent Work in Inductive Logic, *Amer. Philos. Quart.* 1, 249–287, 1964.
17. D. Scott and P. Krauss, Assigning Probabilities to Logical Formulas, in *Aspects of Inductive Logic* (J. Hintikka and P. Suppes, eds.), pp. 219–264. Amsterdam: North-Holland Publ., 1966.
18. P. Krauss, Representation of Symmetric Probability Models, *J. Symbolic Logic* 34, 183–193, 1969.
19. J. Hintikka, A Two-Dimensional Continuum of Inductive Methods, in *Aspects of Inductive Logic* (J. Hintikka and P. Suppes, eds.), pp. 113–132. Amsterdam: North-Holland Publ., 1966.
20. M. Black, *Margins of Precision*, p. 127. Ithaca, New York: Cornell Univ. Press, 1970.
21. W. V. O. Quine, Two Dogmas of Empiricism, reprinted in *From a Logical Point of View*. Cambridge, Massachusetts: Harvard Univ. Press, 1953.
22. R. Carnap and R. C. Jeffrey, eds. *Studies in Inductive Logic and Probability*, I. Berkeley and Los Angeles: Univ. of California Press, 1971.

VIII

Probability as a Pragmatic Necessity: Subjective or Personal Probability

VIIIA. Introduction

The relative-frequency, complexity, classical, and logical interpretations of probability are primarily concerned with knowledge and
inference. None of these interpretations lend themselves to a ready
justification for the use of probability to guide behavior or to facilitate
decision-making. Nor are these interpretations sufficient to direct the
use of probability in decision-making; they require extensive supplementation by *ad hoc* statistical principles. If we focus on the pragmatic
activity of decision-making, as opposed to such epistemic activities as
inference, then it will prove possible to justify and direct the use of a
concept of probability in rational decision-making. The necessity for
a concept of probability in some forms of rational decision-making does
not by itself interpret probability. In the approaches we outline, it will
be seen that the interpretation of probability is related to the interpretation presumed for a preference pattern between decisions. Insofar as
this pattern is subjective or arises from the thought and experience of
an individual, then the resulting probability will also possess a subjective
interpretation.

212

We will outline three versions of the subjective position here and leave the many details, specific axiom systems, and other approaches to the references. Early developments are due to Ramsey [1], de Finetti [2], Savage [3], and Good [4] and relatively recent axiomatic treatments of subjective probability can be found in Fishburn [5, 6], Anscombe and Aumann [7], Pratt *et al.* [8], and Krantz *et al.* [9]. The first approach to subjective probability we discuss follows Anscombe and Aumann in relating subjective judgments to a preexisting (objective) probability scale having a well-defined role in decision-making; it provides a convincing bridge between subjective probability and the approaches we have hiherto discussed. Savage's much discussed work, to the contrary, emphasizes a personalistic decision theory, based only on a primitive notion of preference, that leads to a theory of subjective probability free of all other concepts of probability. Krantz and Luce modify Savage's approach, primarily through a focus on conditional decision rules and secondarily through a different type of axiomatization.

We must first indicate the type of decision problem whose satisfactory resolution requires the introduction of subjective probability. The elements of a decision problem can be taken to be a set $\mathscr{C}$ of consequences, a set $\mathscr{S}$ of states of the world (alternatives), and a set $\mathscr{D}$ of acts or decisions. The set $\mathscr{C}$ can be taken to be either a family of (objective) probability distributions over the field of events generated from all subsets of a finite set of nonrandom payoffs $\{p_i\}$ (the case of a nonrandom consequence is accounted for by the degenerate distribution assigning probability 1 to some payoff) or it can be treated merely as an abstract set. The decisions $d(\cdot) \in \mathscr{D}$ are functions either from $\mathscr{S}$ to $\mathscr{C}$, or, as in the work based on conditional decision rules, from a subset of $\mathscr{S}$ to $\mathscr{C}$. The goal of the decision maker (DM) is to employ whatever knowledge he has about the selection of a state of the world and the values of the consequences so as to select a good decision. Subjective theory suggests criteria of rationality to guard against types of errors, and to improve the easy and reduce the arbitrariness of individual decision-making. It also squarely admits that relevant information about $\mathscr{S}$ may be neither given in the form of a probability distribution nor capable of explicit statement. In this respect the subjective thesis seems more realistic and attractive than that of, say, the relative-frequency school; the latter's purportedly empirical arguments to derive an initial basis for introducing probabilities are far from convincing as sound objective procedures.

In brief, the rationality criteria of the subjectivist are such as to force the DM to act as if he knows a prior distribution σ on $\mathscr{S}$ and a (linear) utility scaling of the consequences; the optimum decision is the one yielding the maximum expected utility—the Bayes principle. While it seems natural to relax the assumptions leading to σ so as to obtain merely a comparative probability ordering of subsets of $\mathscr{S}$, which can in turn yield a representation of a rational preference pattern between decisions, this possibility has not been closely examined (see Section IIG). A brief review of the fundamentals of utility theory will serve to introduce our outline of several approaches to rational decision-making and hence to subjective probability.

VIIIB. Preferences and Utilities

Central to many of the formal theories of subjective probability are arguments whose paradigm is a representation theorem for rational preference patterns over the set $\mathscr{C}$ of consequences when the consequences are objective probability distributions over a sample space of payoffs. The approach taken to the preferencing of consequences, that of utility theory, extends to comparisons of decisions through the introduction of subjective, in lieu of unknown, or meaningless, objective probabilities for the states of the world. We assume that the DM has a preference relation $\lesssim$ between pairs of objective probability distributions, c_1, $c_2 \in \mathscr{C}$, over the sample space of payoffs—"$c_1 \lesssim c_2$" is read "consequence c_1 is not preferred to consequence c_2." We read $c_1 \gtrsim c_2$ as $c_2 \lesssim c_1$. Two consequences c_1 and c_2 are defined to be equivalent, $c_1 \approx c_2$, if $c_1 \lesssim c_2$ and $c_2 \lesssim c_1$. Consequence c_2 is strictly preferred to c_1, $c_1 \prec c_2$, if $c_1 \lesssim c_2$ and it is false that $c_2 \lesssim c_1$. Following Ferguson [10] we present the utility axioms U1–U3 for a "rational" preference relation.

U1. The binary relation $\lesssim$ is a complete order; that is,

(a) $(\forall c_1, c_2 \in \mathscr{C})$ either $c_1 \lesssim c_2$ or $c_2 \lesssim c_1$,
(b) $c_1 \lesssim c_2$ and $c_2 \lesssim c_3$ implies $c_1 \lesssim c_3$.

The requirement of completeness (a) is fairly strong and of questionable acceptability. However, the transitivity requirement (b) seems necessary for a preference relation. If, say, $c_1 \prec c_2$, $c_2 \lesssim c_3$, and $c_3 \lesssim c_1$, then the DM would presumably be willing to pay to be given c_2 in exchange for c_1 and would be happy to exchange c_2 for c_3 and c_3 for c_1. At the end of these exchanges he would return to his starting point of c_1, less the unnecessary initial payment. Nevertheless, some arguments been advanced to defend intransitivity of preferences [11].

Introduce the notation

$$c_3 = \lambda c_1 + (1 - \lambda) c_2 \qquad (0 \leqslant \lambda \leqslant 1)$$

to refer to the consequence c_3 that is the mixture of, or randomization between, consequences c_1 and c_2; that is, if p is a payoff and $c(p)$ the probability of obtaining p with consequence c, then

$$c_3(p) = \lambda c_1(p) + (1 - \lambda) c_2(p).$$

We now assert

U2. $(\forall \lambda > 0)(\forall c_1, c_2, c_3 \in \mathscr{C}) (\lambda c_1 + (1 - \lambda)c_2 < \lambda c_3 + (1 - \lambda)c_2 \Leftrightarrow c_1 < c_3).$

We may think of the mixture $\lambda c_1 + (1 - \lambda) c_2$ as arising from the experiment in which we are awarded c_1 if some coin, with probability λ for heads, falls heads, and we are awarded c_2 if the coin falls tails. If we deny U2, then we are either asserting that we strictly prefer to be given "c_3 on heads and c_2 on tails" to "c_1 on heads and c_2 on tails," even though if heads occur we would not prefer c_3 to c_1, or we are asserting that even though we prefer c_3 to c_1 we would not prefer receiving c_3 rather than c_1 on heads and the same consequence c_2 on tails. Implicit in U2 are three beliefs of somewhat more controversial character. The first is that gambling itself is neither attractive nor repulsive. The second is that $c_3 > c_1$, then no matter how small λ is, we still strictly prefer $\lambda c_3 + (1 - \lambda) c_2$ to $\lambda c_1 + (1 - \lambda) c_2$; there is no region of indifference. The third belief is that the mechanism used in the randomization to produce an event of probability λ is independent of the mechanisms by which consequences may award payoffs. We feel that axioms of the form of U2 are better formulated in terms of particular randomizing experiments than in terms of universal probabilities, as was done in Section IIG.

Finally we present the Archimedean postulate

U3. $(\forall c_1 < c_2 < c_3)(\exists 0 < \mu < \lambda < 1) (\lambda c_1 + (1 - \lambda)c_3 < c_2 < \mu c_1 + (1 - \mu)c_3).$

In part U3 asserts that there are no infinitely good or infinitely disastrous consequences. Perhaps more importantly, U3 assures us that we can always interpolate a strict inequality by a mixture of extremes.

We introduce a

Definition. A function u is a linear utility function agreeing with a preference relation $\lesssim$ on a set $\mathscr{C}$ of consequences if

(a) $u : \mathscr{C} \to R^1$;
(b) $u(c_1) \leqslant u(c_2) \Leftrightarrow c_1 \lesssim c_2$;
(c) $u(\lambda c_1 + (1 - \lambda) c_2) = \lambda u(c_1) + (1 - \lambda) u(c_2).$

We are now able to state

Theorem 1. If $\lesssim$ satisfies U1–U3, then there exists a linear utility function u agreeing with $\lesssim$ on $\mathscr{C}$.

Proof. See Ferguson [10, pp. 14–17]. ∎

Had we strengthened U2 to hold for countable mixtures, then we could have concluded that the utility u must be bounded as well. The implication of the utility theorem is that there is a scaling $u(p)$ of payoffs [$u(p)$ is the utility of the consequence yielding payoff p with probability 1] such that the value of a consequence c is the expected scaled payoff

$$u(c) = \sum_p u(p)\, c(p).$$

VIIIC. An Approach to Subjective Probability through Reference to Preexisting Probability

Both Anscombe and Aumann [7] and Pratt *et al.* [8] develop subjective probability through reference to a preexisting probability scale, although the philosophical position that this scale is objective rather than merely interpersonal is maintained only by Anscombe and Aumann. Our discussion focuses on the treatment of Anscombe and Aumann as it seems to be the most satisfactory to those who firmly believe in at least the partial applicability of objective probability.

1. Axioms of Anscombe and Aumann Type

The notation for the Anscombe–Aumann setting of rational decision making is as follows.

$\mathscr{S} = \{s\}$ finite set of states of nature.

$\mathscr{C} = \{c\}$ set of (objective) probability distributions on some abstract payoff space (also called consequences).

$\gtrsim_\mathscr{C}$ complete ordering of elements of $\mathscr{C}$.

$\mathscr{D} = \{d\}$ set of functions from $\mathscr{S}$ into $\mathscr{C}$.

$(\varLambda, \mathscr{L}, \lambda)$ (objective) probability space presumed to be unrelated to $\mathscr{S}$ and $\mathscr{C}$.

$\mathscr{D}^* = \{\{d_i \mid L_i\}, d_i \in \mathscr{D}, L_i \in \mathscr{L}\}$ set of functions from the finite, measurable partitions $\{L_i\}$ of $\varLambda$ into $\mathscr{D}$; that is,

$$(\forall j)(\forall l \in L_j)(\forall s \in \mathscr{S})\, \{d_i \mid L_i\}(l, s) = d_j(s) \in \mathscr{C}.$$

$\gtrsim_{\mathscr{D}^*}$ complete ordering of elements of $\mathscr{D}^*$.

Note that $\mathscr{D}$ is embedded in $\mathscr{D}^*$ through recourse to the trivial one-set partition of Λ

$$(\forall d \in \mathscr{D}) \{d \mid \Lambda\} \in \mathscr{D}^*.$$

The elements of $\mathscr{D}^*$ are randomized decisions; the experiment $(\Lambda, \mathscr{L}, \lambda)$ is performed and a decision rule $d_i \in \mathscr{D}$ selected if the event $L_i \in \mathscr{L}$ occurs. Corresponding to $d \in \mathscr{D}$ there is a set of consequences of using d,

$$\{c : (\exists s \in \mathscr{S}) \, c = d(s)\}.$$

Corresponding to $\{d_i \mid L_i\} \in \mathscr{D}^*$ there is the set

$$\left\{c : (\exists s \in \mathscr{S}) \, c = \sum_i \lambda(L_i) \, d_i(s)\right\},$$

where $\sum_i \lambda(L_i) \, d_i(s)$ is the consequence assigning payoff p with probability $\sum_i \lambda(L_i) \, d_i(s)[p]$ when $d_i(s)[p]$ is the probability of p under consequence $d_i(s)$.

The axioms we now state for rational decision-making in the preceding framework differ only in technical detail from the axiomatization originally stated by Anscombe and Aumann.

AA1. $\gtrsim_{\mathscr{C}}$ satisfies the utility axioms U1–U3.

AA2. $\gtrsim_{\mathscr{D}^*}$ satisfies U1–U3.

AA3. $(\forall i) \lambda(L_i{}^1) = \lambda(L_i{}^2) \Rightarrow \{d_i \mid L_i{}^1\} \approx_{\mathscr{D}^*} (d_i \mid L_i{}^2\}$.

AA4. $(\forall s \in \mathscr{S}) \, d_1(s) \gtrsim_{\mathscr{C}} d_2(s) \Rightarrow \{d_1 \mid \Lambda\} \gtrsim_{\mathscr{D}^*} \{d_2 \mid \Lambda\}$.

AA5. $c_1 >_{\mathscr{C}} c_2 \, , \, (\forall s) \, d_1(s) = c_1 \, , \, d_2(s) = c_2 \Rightarrow \{d_1 \mid \Lambda\} >_{\mathscr{D}^*} \{d_2 \mid \Lambda\}$.

AA6. $(\forall \{L_i\}$ finite, measurable partition of $\Lambda) \, (\forall d_i \in \mathscr{D}) \sum_i \lambda(L_i) \, d_i \in \mathscr{D}$,

$$\left\{\sum_i \lambda(L_i) \, d_i \mid \Lambda\right\} \approx_{\mathscr{D}^*} \{d_i \mid L_i\}.$$

The justification for AA1 and AA2 depends on the intuitive acceptability of U1–U3. Axiom AA3 asserts that the value of $\{d_i \mid L_i\}$ depends on $\{L_i\}$ only through $\{\lambda(L_i)\}$, the probabilities of the events yielding the decision rules. Axioms AA4 and AA5 are seemingly uncontroversial axioms of dominance or monotonicity of preferences between randomized decisions; AA5 is a special case of the converse to AA4. Axiom AA6 is the key axiom in this approach. In effect AA6 asserts that we are indifferent between performing randomization with respect to $(\Lambda, \mathscr{L}, \lambda)$ either before observing $s \in \mathscr{S}$ [i.e., we first select $d_i(\cdot)$ according to L_i

and then select $d_i(s) \in \mathscr{C}$] or after observing $s \in \mathscr{S}$ [i.e., we first select $\{d_j(s) \in \mathscr{C}\}$ and then select $d_i(s)$ after observing L_i]. Axiom AA6 is only reasonable if we do not expect $(\Lambda, \mathscr{L}, \lambda)$ to interact with the mechanism selecting $s \in \mathscr{S}$ or the particular payoff generated by any given $c \in \mathscr{C}$.

If we adopt these axioms, then we can establish

Theorem 2. If AA1–AA6, then there exist linear utility functions $u_\mathscr{C}$ representing $\gtrsim_\mathscr{C}$ and $u_{\mathscr{D}*}$ representing $\gtrsim_{\mathscr{D}*}$, and they are related through

$$u_{\mathscr{D}*}(\{d_i \mid \Lambda\}) = \sum_{s \in \mathscr{S}} \sigma(s)\, u_\mathscr{C}\big(d(s)\big),$$

where

$$\sigma(s) \geqslant 0, \qquad \sum_{s \in \mathscr{S}} \sigma(s) = 1.$$

Proof. See Ferguson [10, p. 19-21]. ∎

The distribution $\sigma(\cdot)$ is the desired subjective probability scaling our judgments concerning the occurrence of states. Note that we have not exhibited σ explicitly, but only claimed that if one wishes to subjectively compare decisions in which the consequences depend on an "unknown" state of nature, and this comparison is to satisfy certain axioms of rationality, then we must act as if we had assigned a prior distribution σ (based on whatever information or beliefs the DM has) to the states $\mathscr{S}$.

2. The Associated Objective Distribution

The axioms involving the "objective" mixture probabilities $\{\lambda_i\}$ are only reasonable if we properly interpret these probabilities. Why did we assume that a mixture offering a high probability of a desirable consequence is preferable to one offering a low probability for the desirable consequence? The acceptability of this axiom depends on the meaning attributed to probability, and it is not obvious once we free ourselves from implicit belief in as yet unstated views of probability.

One possibility is to assume that the objective probabilities are based on relative-frequency. Assume that we have two consequences $c_1 >_\mathscr{C} c_2$ and two decisions d_1 and d_2, where d_i offers c_1 with relative-frequency-based probability λ_i ($\lambda_2 > \lambda_1$) and c_2 with probability $1 - \lambda_i$. We would then expect that over a large number n of selections of d_i in repeated decision problems, we would achieve c_1 approximately $\lambda_i n$ times and c_2 approximately $(1 - \lambda_i)n$ times. Thus n acceptances of d_i would result in a cumulative fortune F_i of $\lambda_i n$ of c_1 and $(1 - \lambda_i)n$ of c_2. It may now seem reasonable to prefer the fortune F_2 to F_1, as would be true if c_1 and c_2 were monetary payoffs. If this is the case, then d_2 leading to F_2 is preferable to d_1 leading to F_1, as assumed by our axioms.

Unfortunately we did not restrict ourselves to monetary payoffs. As we accumulate consequences our utility for each may change. For example, on a given day we may prefer a good meal to \$1, but prefer a good meal and \$5 to six good meals on the same day. The root of the difficulty with this justification of a relative-frequency interpretation of the probabilities $\{\lambda_i\}$ is that it introduces a new decision problem. This new problem is a compound of n of the original decision problems. We may readily admit that a best decision for the compound problem need not be the n-fold repetition of the best decision for the individual component decision problems.

The above-mentioned difficulty with a relative-frequency-based interpretation can be circumvented by a view of $\{\lambda_i\}$ that is less objective but links more closely with the subjective approach. The axioms that seem reasonable with a relative-frequency interpretation would also seem reasonable if the mixtures were based on the following type of mechanism [8, pp. 272–275]. The DM may be able to find a mechanism generating real numbers $x \in [0, 1]$ such that if d_i offers c_1 if $x \in L_i$ (L_i a subset of $[0, 1]$), and c_2 ($c_1 >_{\mathscr{C}} c_2$) if $x \in \bar{L}_i$ (complement with respect to $[0, 1]$), then d_2 is preferred to d_1 only if the length of L_2 exceeds the length of L_1. We may think of this mechanism as generating a uniformly distributed x, although the meaning of this probability distribution is purely subjective. If we interpret the probabilities $\{\lambda_i\}$ in terms of this mechanism, then the axioms leading to utility and subjective probability seem reasonable, if not virtually circular. That this system is not substantially circular can be seen in the implication that all uncertainties can be quantitatively scaled and are in a sense interchangeable. The uncertainty about extraterrestrial intelligent life existing within 1000 light-years of our solar system is different only in degree and not in kind from the uncertainty about the top card in a well-shuffled deck.

VIIID. Approaches to Subjective Probability through Decision-Making

1. Formulation of Savage

Subjective probability, in this formulation, arises as a factor in a numerical representation for a rational preference ordering of decision functions. The elements of the decision problem are as follows.

$\mathscr{S} = \{s\}$ set of states of the world.

$\gtrsim$ (comparative probability) ordering between subsets, A, B, ..., of $\mathscr{S}$ (events).

$\mathscr{C} = \{c\}$ set of abstract consequences or payoffs.

$\gtrsim_{\mathscr{C}}$ ordering between elements of $\mathscr{C}$.

$\mathscr{D} = \{d(\cdot)\}$ set of functions from $\mathscr{S}$ into $\mathscr{C}$ (decision rules).

$\gtrsim_{\mathscr{D}}$ ordering between elements of $\mathscr{D}$.

We define $\gtrsim_{\mathscr{C}}$ and $\gtrsim$ in terms of $\gtrsim_{\mathscr{D}}$ as follows. Given any $c \in \mathscr{C}$ we denote by $d_c(\cdot)$ the element of $\mathscr{D}$ for which

$$(\forall s \in \mathscr{S}) \, d_c(s) = c.$$

We define

$$c \gtrsim_{\mathscr{C}} c' \Leftrightarrow d_c(\cdot) \gtrsim_{\mathscr{D}} d_{c'}(\cdot).$$

Denote by $d_{c,c'}^A(\cdot)$ the element of $\mathscr{D}$ for which

$$d_{c,c'}^A(s) = \begin{cases} c & \text{if } s \in A \\ c' & \text{if } s \in \bar{A}. \end{cases}$$

We define

$$A \gtrsim B \Leftrightarrow (\forall c, c' \in \mathscr{C}, c >_{\mathscr{C}} c') \, d_{c,c'}^A(\cdot) \gtrsim_{\mathscr{D}} d_{c,c'}^B(\cdot).$$

The reasonableness of these definitions of $\gtrsim_{\mathscr{C}}$ and $\gtrsim$ follows from our expectations that improving the payoffs for fixed events or improving the chances of obtaining the better payoffs will improve the decision.

We require three additional definitions prior to listing Savage's axioms. We say that $f(\cdot)$ and $g(\cdot)$, elements of $\mathscr{D}$, agree on $A \subseteq \mathscr{S}$ if

$$(\forall s \in A) \, f(s) = g(s).$$

We define "$f(\cdot) \gtrsim_{\mathscr{D}} g(\cdot)$ given A" to mean that for all $f'(\cdot)$ agreeing with $f(\cdot)$ on A and $g'(\cdot)$ agreeing with $g(\cdot)$ on A, such that $f'(\cdot)$ and $g'(\cdot)$ agree with each other on $\bar{A}$, $f'(\cdot) \gtrsim_{\mathscr{D}} g'(\cdot)$. Finally we say that $A \subseteq \mathscr{S}$ is null if for all $f(\cdot)$ and $g(\cdot)$, elements of $\mathscr{D}$, $f(\cdot) \gtrsim_{\mathscr{D}} g(\cdot)$ given A. The first two definitions enable us to compare decisions over subsets of their domain. The third definition declares an event to be null if all decisions agreeing on its complement are equivalent.

Savage's axioms for personalistic rational decision-making, renumbered, are as follows.

P1. $(\exists c, c' \in \mathscr{C}) \, c >_{\mathscr{C}} c'.$

P2. $\gtrsim_{\mathscr{D}}$ is a complete order (i.e., complete, transitive, reflexive binary relation).

P3. $(\forall f(\cdot), g(\cdot) \in \mathscr{D})(\forall A \subseteq \mathscr{S})$ either $f(\cdot) \gtrsim_{\mathscr{D}} g(\cdot)$ given A or $g(\cdot) \gtrsim_{\mathscr{D}} f(\cdot)$ given A.

P4. If A is not null, then

$$d_c(\cdot) \gtrsim_{\mathscr{D}} d_{c'}(\cdot) \qquad \text{given} \quad A \Leftrightarrow c \gtrsim_{\mathscr{C}} c'.$$

P5. $(\forall A, B \text{ subsets of } \mathscr{S}) \; A \gtrsim B$ or $B \gtrsim A$.

A finite partition $\mathscr{P}$ of $\mathscr{S}$ is a finite disjoint collection $\{S_i\}$ of subsets of $\mathscr{S}$ whose union is $\mathscr{S}$.

P6. $(\forall f(\cdot) >_{\mathscr{D}} g(\cdot)) \, (\forall c \in \mathscr{C}) \, (\exists \mathscr{P}) \, (\forall S_i \in \mathscr{P})$

$$\left(f_i(s) \equiv \begin{cases} f(s) & \text{if } s \notin S_i, \\ c & \text{if } s \in S_i, \end{cases} \quad g_i(s) \equiv \begin{cases} g(s) & \text{if } s \notin S_i \\ c & \text{if } s \in S_i \end{cases} \right)$$

$$\Rightarrow f_i(\cdot) >_{\mathscr{D}} g(\cdot), f(\cdot) >_{\mathscr{D}} g_i(\cdot).$$

P7. (a) $(\forall s \in A) f(\cdot) \gtrsim_{\mathscr{D}} d_{g(s)}(\cdot)$ given $A \Rightarrow f(\cdot) \gtrsim_{\mathscr{D}} g(\cdot)$ given A.
(b) Replace $\gtrsim_{\mathscr{D}}$ by $\lesssim_{\mathscr{D}}$ in part (a).

Axiom P1 assures us of the nontriviality of the decision problem. If P1 fails, then all decisions are equivalent. P2 asserts that all pairs of decisions are comparable and this, in practice, is an unrealistic assumption. Axiom P3 states that all "conditional" decisions are comparable. Axiom P4 is a version of the sure-thing principle and seems quite reasonable. Axiom P5 asserts that the induced ordering between events is complete. Axioms P1–P5 establish that $\gtrsim$ satisfies the basic axioms C1–C4 of comparative probability. Axiom P6 is the central axiom assuring the existence of a finitely additive P agreeing with $\gtrsim$. As Savage points out, P6 is technically stronger than is necessary to ensure the existence of quantitative subjective probability. However, he finds it an intuitively satisfying postulate. Axiom P7 is an intuitively satisfying statement of dominance that is required in the derivation of representations for $\gtrsim_{\mathscr{D}}$ when $\mathscr{C}$ contains infinitely many consequences; P7 is not needed when $\mathscr{C}$ is finite.

Implicit in our acceptance of P1–P7 is the belief that $\mathscr{S}$ and $\mathscr{C}$ are unrelated entities. That is to say, choices of consequences do not influence the probabilities of events and choices of events do not influence the desirability of consequences. It is not difficult to envision situations in which this unrelatedness is lacking, and the choice of decision affects both the probabilities of states of nature as well as the values of consequences. For example, if in planning a trip I decide to drive rather than to fly, then the state of nature that corresponds to my being delayed by a late flight cannot occur nor would I greatly appreciate the payoff of being served an excellent meal had I flown. These issues are further addressed in the conditional decision formulation of Krantz and Luce discussed in the next subsection.

The key result concerning the existence and role of subjective probability in decision-making is given by

Theorem 3.   Under P1–P7 there exists a unique finitely additive set function (probability) P agreeing with $\gtrsim$, having the property that

$$(\forall A \subseteq \mathscr{S})(\forall 0 \leqslant \rho \leqslant 1)(\exists B \subseteq A)\, P(B) = \rho P(A).$$

Furthermore there exists a real-valued, bounded function u defined on $\mathscr{C}$, unique to within positive linear transformation, such that

$$f(\cdot) \gtrsim_{\mathscr{D}} g(\cdot) \Leftrightarrow Eu(f(s)) \geqslant Eu(g(s)),$$

where expectation is with respect to P.

Proof.   See Fishburn [5, Chapter 14].   ∎

Hence, if we accept P1–P7, then we must choose between decisions as if we could assign a probability to the subsets of $\mathscr{S}$. The theorem says nothing about the interpretation of this probability. However, insofar as $\gtrsim_{\mathscr{D}}$ is subjective in origin (Savage's emphasis on personalistic decision-making), then so will P be subjective in origin. The role of P is clearly not that of making an empirically correct statement about the occurrence of events but is rather that of representing $\gtrsim_{\mathscr{D}}$ as indicated in the theorem. Savage [3] should be consulted for extensive, insightful discussions of his postulates and their implications.

Two formal defects of P1–P7 are the necessity for $\mathscr{S}$ to be uncountable and the requirement that $\gtrsim_{\mathscr{D}}$ be defined for all pairs of decisions. These two defects are somewhat remedied in the formulation of Krantz and Luce.

2. Formulation of Krantz and Luce

Krantz and Luce [9] essentially modify Savage's development of subjective probability by introducing the realistic notion of a conditional decision. A conditional decision is a function from a subset of $\mathscr{S}$ into $\mathscr{C}$. As they point out, there are many instances of decision rules that are only sensibly defined for a subset of the set of possible states of nature;

the choice of decision can delimit the possible states of nature. For example, if the states of nature involve the choices of staying at home or driving elsewhere, then if I decide to stay at home I need not be concerned about the possibility of a collision, a factor that would have been significant had I decided to drive.

Krantz and Luce form the rational decision-making problem as a sextuple $(\mathscr{S}, \mathscr{E}, \mathscr{N}, \mathscr{C}, D, \gtrsim_D)$.

$\mathscr{S}$ set of states of nature.
$\mathscr{E}$ field of subsets of $\mathscr{S}$.
$\mathscr{N}$ subset of $\mathscr{E}$ containing the null events.
$\mathscr{C}$ set of abstract consequences.
D subset of the set of functions from elements of $\mathscr{E}$ into $\mathscr{C}$.
$\gtrsim_D$ ordering of the elements of D.

Elements of $\mathscr{E}$ will be denoted by upper case roman letters (e.g., A, B). Elements of D are the conditional decision functions and they will be denoted by subscripting lower case roman letters with the domain of the function (e.g., $f_A : A \to \mathscr{C}$). To avoid needless repetition in the statement of the axioms we will assume that $f_A, f_A^{(i)}, f_{A \cup R}, g_B, g_B^{(i)}$, and $h_A^{(i)}$ are all in D. We note the convention that

$$(f_A \cup g_B)(s) = \begin{cases} f_A(s) & \text{if } s \in A \\ g_B(s) & \text{if } s \in B. \end{cases}$$

Krantz and Luce proposed and ably discussed the meaning of the following nine axioms of rational decision-making.

KL1.   (i) $\mathscr{E} - \mathscr{N}$ has at least three pairwise disjoint elements.
(ii) $D/\approx_D$ has at least two distinct equivalence classes.

KL2.   (i)   If $A \cap B = \varnothing$, then $f_A \cup g_B \in D$.
(ii)   If $B \subset A$, then the restriction of f_A to B is in D.

KL3.   (i)   If A and g_B are given, then there exists $h_A \in D$ for which $h_A \approx_D g_B$.
(ii)   If $A \cap B = \varnothing$ and $h_A^{(1)} \cup g_B \gtrsim_D f_{A \cup B} \gtrsim_D h_A^{(2)} \cup g_B$, then there exists $h_A \in D$ such that $h_A \cup g_B \approx_D f_{A \cup B}$.

KL4.   (i)   If $R \in \mathscr{N}$ and $S \subset R$, then $S \in \mathscr{N}$.
(ii) $R \in \mathscr{N}$ iff, for all $f_{A \cup R} \in D$ with $A \cap R = \varnothing$, $f_{A \cup R} \approx_D f_A$, where f_A is the restriction of $f_{A \cup R}$ to A.

KL5. $\gtrsim_D$ is a complete ordering of D.

KL6.   If $A \cap B = \varnothing$ and $f_A \approx_D g_B$, then $f_A \cup g_B \approx_D f_A$.

KL7.   If $A \cap B = \varnothing$, then $f_A^{(1)} \succsim_D f_A^{(2)} \Leftrightarrow f_A^{(1)} \cup g_B \succsim_D f_A^{(2)} \cup g_B$.

Before stating axioms KL8 and KL9, we need to define a standard sequence of conditional decisions $\{f_A^{(i)}, i \in N\}$.

Definition. $\{f_A^{(i)}, i \in N\}$ is a standard sequence of conditional decisions if for some $B \in \mathscr{E} - \mathscr{N}$, $A \cap B = \varnothing$, $g_B^{(0)}, g_B^{(1)} \in D$ with false $g_B^{(0)} \approx_D g_B^{(1)}$, then

$$(\forall i, i + 1 \in N) f_A^{(i)} \cup g_B^{(1)} \approx_D f_A^{(i+1)} \cup g_B^{(0)}.$$

KL8.   Any strictly bounded standard sequence is finite.

KL9.   If $\{f_A^{(i)}, i \in N\}$ and $\{h_B^{(i)}, i \in N\}$ are any two standard sequences such that, for some j, $j + 1 \in N$, $f_A^{(j)} \approx_D h_B^{(j)}$ and $f_A^{(j+1)} \approx_D h_B^{(j+1)}$, then for all $i \in N$, $f_A^{(i)} \approx_D h_B^{(i)}$.

The first three axioms are structural in that they require $\mathscr{E} - \mathscr{N}$ not to be too small and to satisfy certain closure properties. Axioms KL1–KL3 are not consequences of reflections on the meaning of rational procedures. While KL1 is unobjectionable, the same cannot be said for KL2 and KL3. In fact KL2 somewhat contradicts the motivation behind the introduction of conditional decisions as a more realistic model than that of unconditional decisions. If A is the event that I remain at home and B is the event that I drive somewhere, and f_A some course of action at home and g_B some choice of driving route, then it is not very reasonable to require me to have preferences concerning the unrealistic choice $f_A \cup g_B$. Perhaps the best defense of KL2 is that it yields a theory of subjective probability and rational decision-making which uses weaker assumptions than the other theories we have mentioned. The unconditional decision theories, it can be argued, implicitly commit the DM to such decisions and more, but hide it better. We are also concerned about KL3, as we feel that solvability assumptions in terms of equivalences rather than in terms of indifference are in general unrealistic demands. Albeit, such requirements are rife in most formulations of decision theory which yield optimal decisions.

The remaining six axioms are constraints imposed by an intuitive notion of a rational decision structure. Axiom KL4 essentially defines what is meant by a null event; we can ignore consequences of the occurrence of any subset of a null event. Axiom KL5 can be criticized for requiring a complete ordering of D as well as for not admitting judgments of indifference (symmetric, reflexive, intransitive). KL6 adds substance to the interpretation of the conditionality of conditional decisions. If A occurs, then $f_A \cup g_B$ has consequences identical to f_A, whereas if B occurs, then $f_A \cup g_B$ has consequences g_B that are equivalent to those of f_A. Axiom KL7 is a version of the sure-thing principle. Restated, if consequences f_A and g_A agree on $B \subset A$, then preferences between them are determined by their restrictions, f_{A-B} and g_{A-B}, to $A - B$ where they may disagree. Both KL8 and KL9 assert reasonable properties for standard sequences of consequences. Axiom KL8 asserts that there are neither infinitesimally nor infinitely preferable consequences; KL9 affirms that standard sequences or series have the desired linear scale property in that two linear functions $\{f_A^{(i)}\}$ and $\{h_B^{(i)}\}$, agreeing at two points are identical.

The implications of KL1–KL9 for rational decision-making via conditional expected utility are given in

Theorem 4. If KL1–KL9, then there exist a finitely additive probability P on $\mathscr{E}$ and a real-valued function u on D such that for all A, $B \in \mathscr{E} - \mathscr{N}$, f_A, $g_B \in D$,

    (i) $R \in \mathscr{N} \Leftrightarrow P(R) = 0$.

    (ii) $f_A \gtrsim_D g_B \Leftrightarrow u(f_A) \geqslant u(g_B)$.

    (iii) $A \cap B = \varnothing \Rightarrow u(f_A \cup g_B) = u(f_A)P(A \mid A \cup B) + u(g_B)P(B \mid A \cup B)$.

Furthermore, P is unique and u is unique to within a positive linear transformation.

Proof. See Krantz *et al.* [9, Sect. 8.3]. ∎

The distribution P whose existence is implied by KL1–KL9 is the desired subjective distribution over the states of nature when $\gtrsim_D$ is assumed to be subjective in origin. A DM who accepts KL1–KL9 must make decisions as if he knew the prior probability of subsets of the states of nature. For him to fail to act in this manner would imply behavior contradicting one or more of these reasonable axioms.)

The advantages of the Krantz–Luce formulation over the earlier one of Savage include:

(1) The state space can be finite in Krantz–Luce although not in Savage.

(2) The set of conditional decisions is not required to be complete in Krantz–Luce, whereas all (unconditional) decisions are included in Savage; in particular the former theory does not force the inclusion of constant decision functions.

However, inspection of Savage's axioms, and his discussion of them, suggests that in an attempt at gaining intuitive plausibility, he did not attempt to postulate the mathematical minimum needed for an expected utility representation of a rational preference between decisions. Furthermore, Savage managed to avoid assumptions similar to KL3 and KL6. Both theories can lay claims to being compelling, although neither is completely persuasive.

VIIIE. Subjective versus Arbitrary: Learning from Experience

While assessments of subjective probability are personal, they need not be completely arbitrary. The axioms establishing the existence of subjective probability allow the DM great freedom in making his introspective assessments based on whatever information he possesses. However, as he gains information through further experience we would expect him to reassess his earlier judgments in a reasonable manner. A model of the way in which a rational DM adjusts his probability assessments after receipt of explicit additional information can be based on a concept of subjective conditional probability.

The DM's initial assessment of the subjective probabilities of subsets of $\mathscr{S}$ is given by $\sigma(\cdot)$. After the DM receives information that leads him to believe that the true state is in $E \subseteq \mathscr{S}$, he readjusts his subjective probabilities to the new nonnegative, unit-normed, finitely additive set function $\sigma(\cdot \mid E)$. The relation between $\sigma(\cdot \mid E)$ and $\sigma(\cdot)$ is to be determined from their respective roles in decision-making. The role of $\sigma(\cdot \mid E)$ is to be that by evaluating the conditional expected utility of a decision rule

$$u_{\mathscr{D}}(d \mid E) = \sum_{s \in S} u_{\mathscr{C}}(d(s)) \, \sigma(s \mid E)$$

(the use of summation is intended to skirt technicalities of integration), the DM can determine the preferred decision *given* that he believes E contains the true state.

If E contains the true state, then the values of d on $\bar{E}$ should be irrelevant for our preferences. Hence if the functions d, d' agree on E we should find that

$$u_{\mathscr{D}}(d \mid E) = u_{\mathscr{D}}(d' \mid E) \quad \text{or} \quad \sum_{s \in \bar{E}} \sigma(s \mid E)[u_{\mathscr{C}}(d(s)) - u_{\mathscr{C}}(d'(s))] = 0.$$

It follows that

$$\sigma(\bar{E} \mid E) = 0.$$

If, on the other hand, d, d' agree on $\bar{E}$, then we would expect that

$$u_{\mathcal{D}}(d) \geqslant u_{\mathcal{D}}(d') \Leftrightarrow u_{\mathcal{D}}(d \mid E) \geqslant u_{\mathcal{D}}(d' \mid E);$$

if they agree on $\bar{E}$, then our (unconditional) preferences between them depend only on their behavior on E. It follows from

$$\sum_{s \in E} \sigma(s)[u_{\mathcal{G}}(d(s)) - u_{\mathcal{G}}(d'(s))] \geqslant 0 \Leftrightarrow \sum_{s \in E} \sigma(s \mid E)[u_{\mathcal{G}}(d(s)) - u_{\mathcal{G}}(d'(s))] \geqslant 0,$$

that

$$(\exists \alpha > 0)(\forall s \in E)\, \sigma(s \mid E) = \alpha\sigma(s).$$

The requirement that $\sigma(\cdot \mid E)$ itself be a subjective probability assignment yields

$$\alpha = 1/\sigma(E).$$

Summing up the preceding argument we see that

$$(\forall A, E \subseteq \mathcal{S})\, \sigma(A \mid E) = \frac{\sigma(A \cap E)}{\sigma(E)},$$

the usual equation for conditional probability. Hence the rational DM has little option in the way he adjusts his subjective probability assessments after receipt of explicit information that leads him to believe the true state lies in some subset of $\mathcal{S}$.

What though of information, such as the outcomes of previous experiments, that does not seem to inform the DM as to a subset of $\mathcal{S}$ within which the true state lies? A typical example is one in which the DM believes that one of a parametric family of distributions $\{P_\theta, \theta \in \Theta\}$ describes the random observation $\chi \in X$. Prior to observing χ the DM assigns a subjective prior π to subsets of Θ. How should he then rationally modify $\pi(\cdot)$ to $\pi(\cdot \mid \chi)$ after observing χ?

This problem reduces to the one we have treated if we consider the enlarged set of states

$$\mathcal{S} = \Theta \times X,$$

and note that the DM's belief in the relevance of the parametric model means that his subjective prior σ on $\mathcal{S}$ is of the form (assuming X discrete for convenience)

$$\sigma(\theta, \chi) = \pi(\theta)\, P_\theta(\chi).$$

Observation of χ now permits us to say that the true state lies in

$$E = \{(\theta, \chi') : \chi' = \chi\} \subseteq \mathcal{S}.$$

We can now directly apply the preceding discussion to conclude that

$$\pi(\theta \mid \chi) = \frac{P_\theta(\chi)\,\pi(\theta)}{\sum_{\theta' \in \Theta} P_\theta(\chi)\,_\pi(\theta')},$$

as expected.

Thus we can generally expect that when the DM correctly believes a sequence of experiments to be independent and identically distributed or exchangeable with the distributions known only to lie in some parametric family, then as his information $(\chi_1, \ldots, \chi_n)$ about past outcomes increases, his subjective probabilities $\pi(\cdot \mid \chi_1, \ldots, \chi_n)$ will converge in some probabilistic sense to a distribution degenerate at the true parameter. The validity of this conclusion, of course, requires regularity conditions, but, more importantly, the initial assessment $\pi(\theta)$ must have assigned positive probability to the true parameter. It is through such arguments, developed more fully in Pratt *et al.* [8, Chapter II], Savage [3, pp. 46–55], and in de Finetti [2, pp. 142–147], that a subjective probabilist comes to exhibit some degree of behavioral conformity with the expectations of an "objective" relative-frequentist.

On the negative side it can be argued [12] that the preceding analysis of how we learn from experience enormously oversimplifies the wealth and complexity of our experience. How do we decide which information to process out of the enormous surfeit of impressions? Furthermore, is it not inconsistent with the subjectivist philosophy to attempt to account completely for the time evolution of subjective probability in terms of explicit calculations with explicit data? The subjectivists have opened a door to subjective assessments without really seeming to know how to narrow that opening properly.

VIIIF. Measurement of Subjective Probability

When $\mathscr{S}$ is a small finite set there is a straightforward way to measure a DM's subjective probability distribution σ that is suggested by the formal approach of Section VIIIC. Let the DM select two consequences c and c' with $c \succ_{\mathscr{C}} c'$. Define the decision functions

$$d^A(s) = \begin{cases} c & \text{if } s \in A \\ c' & \text{if } s \notin A, \end{cases}$$

$$d_\lambda(s) = \lambda c + (1 - \lambda)\,c',$$

where $0 \leqslant \lambda \leqslant 1$ is an "objective" probability. The DM is then to evaluate

$$\lambda^* = \sup\{\lambda : d_\lambda \lesssim_{\mathscr{D}} d^A\}$$

by introspective consideration of his preferences for d_λ versus d^A. It follows from the formulation of subjective probability in Section VIIIC that

$$\lambda^* = \sigma(A).$$

In principle this exercise in introspection is repeated for each subset of $\mathscr{S}$, although it suffices, by the additivity of σ, to evaluate σ for the atoms of $\mathscr{S}$.

Somewhat more realistic attempts at measurement of subjective probability are discussed in Savage [13]. The key to these attempts is the provision of an actual decision-making situation wherein there are payoffs depending on the reported subjective probabilities. If for example we assign a payoff $p(\hat{\sigma}(s))$ to an estimate $\hat{\sigma}(s)$ of $\sigma(s)$ and assume (improperly) that the decision maker wishes to maximize his expected payoff, $\sum_s \sigma(s) p(\hat{\sigma}(s))$, then taking

$$p(x) = \log(x),$$

will assure that the maximizing $\hat{\sigma}$ is σ. The imperfection in this scheme is of course that the decision maker at best maximizes his expected utility and not his expected payoff.

When $\mathscr{S}$ is not a small finite set then these methods of exhaustive evaluation are completely impractical. When $\mathscr{S}$ is a subset of R^n, approximate methods have been suggested [14, 15]. First the DM selects one of a set of standard functional forms for σ. He then evaluates points in the distribution from introspective determinations of σ for selected subsets of $\mathscr{S}$ and "fits" the standard form to σ. The interested reader can find details in the references.

When $\mathscr{S}$ has the cardinality of a function space, as it might if we attempted to use subjective probability in a nonparametric statistical problem, then there are very few ideas as to how to proceed.

VIIIG. Roles for Subjective Probability

Why should we bother to determine a subjective probability distribution, utility function, and a numerical representation of the preference pattern over the set of decisions when we assume that the DM, who will use the representation, already knows the pattern? Perhaps the chief purpose behind this circuitous proceeding is to enable the DM to replace a complicated decision problem by a set of simple ones. From the DM's consistent and confident answers to a set of simple, generally dichotomous, decision problems we can extract his σ and u. These results can then be applied to evaluations of complicated acts or decisions, between which the DM may be uncertain in his preferences, so as to yield a decision consistent with the DM's more strongly held preferences. Furthermore, since the complex decision is selected through the maximization of expected utility, the DM is assured that his decision is coherent (see Section VIID). The axioms of subjective probability, at the least, are intended to free decision-making from some forms of error.

Practical illustrations of the advantages of reducing complex decision problems to a set of simpler ones through the use of subjective probability seem to be most prevalent in the area of business administration and management [15].

A secondary ground for interest in σ is that it enables us to communicate, or make interpersonal, personal judgments about the state of the world. Perhaps DM_1 has to make a decision about the world with states $\mathscr{S}$ but believes himself to be very uninformed about $\mathscr{S}$ in comparison with DM_2. If DM_1 and DM_2 accept the axioms of rational decision-making underlying the theory of subjective probability and if they agree as to what constitutes an "objective" probability, then DM_2 can communicate his "expert" judgments about $\mathscr{S}$ to DM_1 through specification of σ. This communication will be meaningful and useful even though DM_1 and DM_2 may disagree in their evaluations of consequences. For example, the doctor can announce his subjective probabilities for the states of health of the patient after various treatments. The patient can then combine the doctor's assessments of probabilities with his own evaluations of the consequences of the outcomes of the possible treatments and thereby select the "best" treatment.

VIIIH. Critique of Subjective Probability

We present our criticisms as they apply to the roles, formulation, measurement, and justification of subjective probability. Many of the objections we raise have previously been considered by Savage.

1. Role of Subjective Probability

Not all of our inductive behavior leads directly to a decision. We often collect information and learn without specific ends in view—how else would one characterize research and science? Probability and statistics have a long established role in the summarization and reduction of data and the construction of inferences—in the derivation of conclusions rather than decisions. We need not repeat here the defense of the rights of conclusions versus decisions by Tukey [16]. Yet in the absence of a decision problem there is no obvious preference relation from which to derive probabilities, and the subjective approaches discussed in this chapter are unable to lead to conclusions. As Savage[†] [17] has put it, "The philosophical puzzle is how the theory can bear on situations in which any notion of motive seems inapplicable."

Subjective probability lies as the root of what is known as the Bayesian school of statistics. The attractiveness of this school may stem from the relative simplicity of the resulting statistical theory once distributions are provided for all parameters. However, Bayesian statistics can provide a false sense of security. It is neither rational nor wise to force what few crumbs of information we may possess about a parameter into the misleadingly detailed form of a distribution. It is a seeming defect of most of the current theories of subjective probability that they do not distinguish between those prior distributions arrived at after much experience and those based on only the vaguest forms of uninformed introspection. (One exception to this criticism is the approach of Smith [18], where allowance is made for ranges of subjective probabilities.) We may be much better off selecting other decision criteria, after careful thought as to what we are trying to achieve with our statistical design, than if we insist on casting all problems in the same, albeit attractive, mold.

2. Formulation of Subjective Probability

In the formulations of Savage and Krantz and Luce subjective probability rests completely on the primitive concept of a preference relation between decisions. However, the notion of preference is not one with which we are intuitively fully comfortable. Where does our knowledge of these preferences come from and what justifies them? Von Wright [19] has even argued that probability may be logically prior to preference. In his view we determine reasoned preferences only after having previously assessed probabilities. While he grants a form of unreasoned preference that is not based on assessments of probability, he does not believe that such unfounded and unstable judgments should be the basis of a subjective theory. Perhaps fact (probability of occurrence

---

[†] L. J. Savage, Implications of Personal Probability for Induction, *J. Philosophy* **LXIV** 599, 1967.

of states of nature) and value (preferences for payoffs) are not as independent as they usually seem to be.

What licenses the personal idiosyncrasies countenanced by the various theories of subjective probability? At first glance it would seem that individuals relevantly differ only in their values, experiences, and abilities to reason. Disagreements in valuation should be fully accounted for by personal utility functions for consequences and should not affect the assessments of subjective probabilities. With regard to differing experiences or bodies of relevant information, Edwards *et al.* [20] have expressed doubts that this is the only legitimate basis of a difference of opinion. Apparently, subjective theories would permit disagreement between individuals sharing the same relevant experience. Admittedly the notion of two individuals sharing the same relevant experience is highly unrealistic in most situations.

If we control for differences in values and in experiences, then it appears that only differences in ability to "reason" can account for the discrepancies in subjective probability assessments. Either there is a unique correct way to reason from experience to probability or there is not. If there is a unique correct way, then the subjectivist position is in error, for it countenances disagreements between individuals due to errors in reasoning. If, as seems more realistic, there is no unique correct way of reasoning, then there is no reason to favor one possible subjective assessment over another. If this is the case, then it seems better to recognize that probability is not quantitative in the usual sense than to insist upon obtaining a spurious quantitative probability; *ex nihilo nihil* (out of nothing, nothing). We do not agree with de Finetti and Savage [21] that "attempts to say that exact probabilities are 'meaningless' or 'nonexistent' pose more severe problems than they are intended to resolve."

A possible advantage of the subjective approach is that, by relying on raw experience that is not necessarily expressed in a language, we circumvent the usual procedures of converting from "raw experience" to language, reasoning in the language, and then converting linguistic conclusions into decisions intended to influence the world of experience. In the subjective approach experience converts "directly" into decisions, thereby avoiding problems of language. Language expresses universals, whereas raw experience and actual situations are more nearly particulars. Perhaps by avoiding unreliable, approximate conversions between universals and particulars the subjectivist can make better judgments than can his objectivist competitors. However, the subjectivist, while properly opening the door to this possibility, seems unable to control the entrance of totally extraneous factors.

In a more formal vein, the axiomatic requirement that a rational preference pattern between decisions is complete (all decisions are comparable) is common to most formulations of subjective probability. This assumption, is, however, at odds with the actual behavior of a DM. Introspection reveals that we are not always able to preference between decisions among which we are not indifferent. As Wolfowitz [22] put it, "Why, when there is so much debate over the choice of one [he refers here to the best rule], require the statistician to order all?" The requirement that the preferencing be complete is not a requirement forced upon us by considerations of rationality. It is a mathematical convenience reflective of an idealized view of a DM. The Krantz–Luce formulation has the advantage in this regard of not requiring consideration of all possible decision rules. It seems likely that future formulations of rational decision-making will require even less by way of comparisons between elements of a large set of decisions and may yield subjective comparative probabilities in place of less realistic quantitative probabilities.

3. Measurement of Subjective Probability

A practical utilization of the idea of subjective probability requires a reasonable method for extracting the subjective probabilities from the individual. Naive methods of assessment may well yield results clouded by psychological irrelevancy and misunderstanding. The problem of the measurement of utility and subjective probability has been discussed in several sources [13, 23]. Difficulties are inherent in the fact that subjective probability is an oversimplified theory of individual decision-making. Its program cannot be carried out if human DMs are unable or unwilling to establish preference patterns between decisions as envisaged in the theory. We will only mention three of the many problems that beset measurement of subjective probability.

The axioms are only those of consistent decision and cannot of themselves correct inconsistencies in a DM's assessments. Three sources of inconsistency are a lack of transitivity in a stated preference pattern, violations of the sure-thing principle, and fluctuations in the preferences asserted between inequivalent consequences. A violation of transitivity can easily occur in a sufficiently long sequence of comparisons, and the axioms of subjective probability provide no guide to the resolution of such an inconsistency. Albeit, when such violations are noticed by DMs they are willing to attempt to remove them. Violations of the sure-thing principle appear to be more serious. Even when it is pointed out to them, many DMs will not change the preferences that yield these violations [24]. Finally, it has been observed that a DM may change his revealed preferences over a short time period during which he has to all appearances received no new information. Block and Marschak [25] and others have proposed to deal with this difficulty by introducing the notion of stochastic preferences; for example, the consequence c is "stochastically preferred" to c' if the empirical probability with which c is selected over c' in repeated opportunities is greater than $1/2$. At a minimum, these remarks indicate a restricted role in the domain of decision problems for the theories of subjective probability we have outlined. We are enabled to recognize rational decision-making without being told how to do it.

A second source of measurement problems stems from the difficulty of recalling all that may be relevant to some problem. The early treatments of the subjective approach firmly urged the DM to weigh all available information in making his assessments. However, later reflection seems to have led to acceptance of the facts that the DM can neither recall all of his information on demand nor is he willing or able to think hard enough about half-remembered facts or about the truth or falsity of logical propositions (theorems). Savage [26] and Hacking [27] have both considered this problem, with Savage hesitantly speculating on the possibility of introducing a cost for thinking.

While subjective probability claims to be part of a normative theory of decision-making, and not a descriptive theory of psychological processes, it can deviate greatly from a descriptive theory only at the peril of being inapplicable.

4. Justification of Subjective Probability

As discussed in Section IB, we may justify a theory by showing that it is as least as good as all other theories of probability or we may look for pragmatic grounds on which to vindicate it. We first discuss the claims advanced for the primacy of subjective probability, and then consider the pragmatic arguments of avoidance of error and the achievement of self-satisfaction.

It has been suggested that subjective probability subsumes all other approaches to probability. We have already indicated in Section VIIIE how a subjectivist might behave indistinguishably from a relative-frequentist as he learns from experience. With respect to classical probability, it has been argued that at root the recognition of symmetry and equiprobable alternatives is a subjective phenomenon. In those situations where the subjectivist believes there is symmetry he would act as if he accepted classical probability, for example, the arguments leading to the Laplacian prior presented in Chapter VI. Carnap, in seeming agreement with this position, tried to orient logical probability to a position of compatibility with subjective probability or at least with rational decision-making. Should the subjectivist in his appraisal of a decision problem be led to accept, say, Carnap's axioms for logical probability, then he would act in accordance with the prescription of logical probability. Indeed, notwithstanding Carnap's attempt to found acceptance of his axioms on an *a priori* basis, it is likely that those who use his theory have a pronouncedly subjective basis for doing so. Perhaps, say, the choice of λ in C_λ is best made subjectively.

While these observations seem to support the general subjectivist position, they do not directly support any of the several formulations of subjective probability. Furthermore, with respect to the general position, while as thinkers we are inevitably subjective, all thought is not. It may be true that we have to give our individual assent to any theory of probability, but this does not rule out the development of some approach to probability so compeling that henceforth all "rational" individuals will assess and use probability in the same objective way.

Turning to the vindication of subjective probability, we see that by requiring the DM to act in accordance with the maximization of expected utility we are assuring that his decision-making will be coherent. The virtue of coherence is that it eliminates decisions by which the DM may lose but cannot possibly win. Hence, by following the dictates of sub-jective probability the DM can avoid certain errors. This protection afforded by coherence may in some instances also be achievable in other ways. For example, consider the two wagers W_1 and W_2 on the outcome of a horse race given by:

(W_1) I pay you \$2 if horse A wins and you pay me \$1 if A loses.
(W_2) I pay you \$2 if horse A loses and you pay me \$1 if A wins.

Clearly by offering both W_1 and W_2, I am not coherent; if you accept both wagers, then I am assured a net loss of \$1. However, if as is possible I only allow you to select one of W_1, W_2, then the outcome is not determined. In other words, I may be indifferent between two decisions, d_1 and d_2, yet strictly prefer either to a mixture,

$$\lambda d_1 + (1 - \lambda) d_2 .$$

The avoidance of error is only desirable when it is not at the expense of incurring greater errors. As we have noted before, there may not exist procedures for inference or decision-making which have all of the intuitively desirable properties. If that is the case, then avoidance of error becomes a relative and not an absolute matter.

Curiously, the argument that there may not exist a best rational procedure of decision-making can be turned to the advantage of subjective probability. Of all the methods for decision-making, that based on subjective probability is most likely to satisfy its user. Within limits set by the ability of the DM to conform to the prescription of subjective probability, he is encouraged by this approach to make decisions that agree with his preferences as evaluated to the best of his personal knowledge and belief. In the absence of a unique best set of instructions for rationally reasoning from knowledge and beliefs to preferences, self-satisfaction is the best that could be expected.

References

1. F. Ramsey, Truth and Probability, reprinted in *Studies in Subjective Probability* (H. Kyburg and H. Smokler, eds.), pp. 61–92. New York: Wiley, 1964.
2. B. de Finetti, Foresight: Its Logical Laws, Its Subjective Sources, reprinted in *Studies in Subjective Probability* (H. Kyburg and H. Smokler, eds.), pp. 93–158. New York: Wiley, 1964.
3. L. J. Savage, *The Foundations of Statistics*. New York: Wiley, 1954.
4. I. J. Good, *Probability and the Weighing of Evidence*. London: Griffin, 1950.
5. P. C. Fishburn, *Utility Theory for Decision Making*. New York: Wiley, 1970.
6. P. C. Fishburn, *Decision and Value Theory*. New York: Wiley, 1964.
7. F. Anscombe and R. Aumann, A Definition of Subjective Probability, *Ann. Math. Statist.* **34**, 199–205, 1963.
8. J. Pratt, H. Raiffa, and R. Schlaifer, *Introduction to Statistical Decision Theory*. New York: McGraw-Hill, 1965.
9. D. Krantz, R. D. Luce, P. Suppes, and A. Tversky, *Foundations of Measurement*, Vol. I, Chapter 8. New York: Academic Press, 1971.
10. T. Ferguson, *Mathematical Statistics*, pp. 12–21. New York: Academic Press, 1967.
11. A. Tversky, Intransitivity of Preferences, *Psychol. Rev.* **76**, 31–48, 1969.
12. J. Hintikka and P. Suppes, eds., *Aspects of Inductive Logic*, pp. 49–65. Amsterdam: North-Holland Publ., 1966.
13. L. J. Savage, The Elicitation of Personal Probabilities and Expectations. Dept. of Statist., Yale Univ., New Haven, Connecticut, October 13, 1970.
14. R. Meyer and J. Pratt, The Consistent Assessment and Fairing of Preference Functions, *IEEE Trans. Systems Sci. Cybernetics* **SSC-4**, 270–278, 1968.
15. R. Schlaifer, *Analysis of Decisions Under Uncertainty*, pp. 140–170, 282–305. New York: McGraw-Hill, 1969.
16. J. Tukey, Conclusions *vs.* Decisions, *Technometrics* **2**, 423–433, 1960.
17. L. J. Savage, Implications of Personal Probability for Induction, *J. Philosophy* **LXIV**, 599, 1967.
18. C. A. B. Smith, Consistency in Statistical Inference and Decision, *J. Roy. Statist. Soc. Ser. B* **23**, 1–37, 1966.

19. G. H. von Wright, Remarks on the Epistemology of Subjective Probability, in *Logic, Methodology and Philosophy of Science* (E. Nagel, P. Suppes, and A. Tarski, eds.), pp. 330–339. Stanford, California: Stanford Univ. Press, 1962.

20. W. Edwards, H. Lindeman, and L. J. Savage, Bayesian Statistical Inference for Psychological Research, *Psychol. Rev.* **70**, 197, 1963.

21. B. de Finetti and L. J. Savage, On the Manner of Choosing Initial Probabilities, p. 4. Summary prepared by L. J. Savage, Yale University, New Haven, Connecticut, February 15, 1962.

22. J. Wolfowitz, Bayesian Inference and Axioms of Consistent Decision, *Econometrica* **30**, 476, 1962.

23. D. Davidson, P. Suppes, and S. Siegel, *Decision-Making an Experimental Approach.* Stanford, California: Stanford Univ. Press, 1957.

24. K. MacCrimmon, Descriptive and Normative Implications of the Decision Theory Postulates, in *Risk and Uncertainty* (K. Borch and J. Mossin, eds.), pp. 3–32. New York: Macmillan, 1968.

25. H. Block and J. Marschak, Random Orderings, Cowles Foundation Discussion Paper 42. Yale Univ., New Haven, Connecticut, 1957.

26. L. J. Savage, Difficulties in the Theory of Personal Probability, *Philos. Sci.* **34**, 308, 1967.

27. I. Hacking, Slightly More Realistic Personal Probability, *Philos. Sci.* **34**, 311–325, 1967.

IX

Conclusions

IXA. Where Do We Stand?

1. With Respect to Definitions of Probability

Let us summarize the status of the various theories of quantitative probability that we have discussed with respect to their ability to measure probability and to yield probability conclusions of value either for the characterization of chance and uncertainty or for the determination of inferences and decisions.

The formal theories of comparative and quantitative probability treated in Chapters II and III are insufficient guides to either the measurement of probability or to the suitable role for probability conclusions. While the axioms limit the possibilities for interpretation there are still, as we have seen, too many very different interpretations (e.g., mass or length) compatible with the same formal structure. Of course, the more one can supply by way of interesting definitions (e.g., independence, random variable, etc.), the more tightly circumscribed are the compatible interpretations. However, the formal theories are presently far from adequate guides to practice.

The advantage of a finite relative-frequency interpretation of probability is that it easily answers the measurement question, at least for

238

those random phenomena that are unlinkedly repeatable. The grave disadvantages are the absence of strong grounds for relying on such probabilities for inference or decision-making and the narrow scope of such an interpretation. Finite relative-frequency is descriptive of the past behavior of an experiment but is difficult to justify as being predictive of future behavior. Our experience is that the frequency in the first N trials is usually not the frequency in the next N trials. Furthermore, not all random phenomena are amenable to analysis in terms of arbitrary repetitions. Why are not unique occurrences fit for probabilistic analysis ? If we look closely, we see that we never exactly repeat any experiment. Nevertheless, it appears that informal recognitions of approximate repetitions coupled with a finite relative-frequency interpretation is the most commonly applied theory of probability.

As discussed in Chapter IV, we find little to say in favor of a limit of relative-frequency interpretation of probability either as to its empirical origins or in its formulations by von Mises, through the collective, or by Bernoulli/Borel, through the laws of large numbers. A limit interpretation is of value neither for the measurement of probability nor for the application of probability. A limit statement without rates of convergence is an idealization that is unlike most of the idealizations in science. In science, while we do not make exact measurements, we at least expect to achieve some known degree of proximity (e.g., the speed of light is not less than 185,000 mps). Knowing the value of the limit without knowing how it is approached does not assist us in arriving at inferences. The relative-frequency-based theories are inadequate characterizations of chance.

The theory of computational-complexity-based probability, treated in Chapter V, while successful at categorizing sequences, is as yet insufficiently developed with respect to the concept of probability. However, the indications are that when developed, this approach will be able to measure probability, but will encounter difficulties with the justification of the use of the measured probabilities. The justification problem seems to be very similar to that faced by other logical theories of probability.

Classical probability, as noted in Chapter VI, is ambiguous as to the grounds for and methods of assessment of probability. It partakes of elements of the logical and subjective concepts and is far less clear than the logical theory as to how to reach probability assessments. It is also perhaps true that the subjectivist claim to subsume classical probability as a special case is valid. In the absence of a clear interpretation of classical probability we cannot arrive at a determination of a justifiable role for it. Of course, classical probability is often invoked to supply uniform prior probabilities as a characterization of "complete" prior ignorance, but this is an inconsistent and dubious practice. The axiomatic reformulations remove some of the measurement ambiguities but do little to advance the problem of justification.

Quantitative logical probability as conceived of by Carnap and discussed in Section VIID and following is as yet incompletely developed. The axiom system that leads to the C_λ or λ system is *ad hoc* and not clearly preferable to other axioms that would lead to C^* or other confirmation function. Thus the measurement problem for logical probability does not seem to have been satisfactorily solved. What is more serious is the selection of a role for Carnapian probability. Formal processing of empirical statements need not lead to empirically valid conclusions. Logical probability does have promise as an explication of rational inference making or inductive reasoning, although the practical value of "rationality" has been challenged. (Perhaps such challenges are ill-founded.) Carnap, as we have noted, attempted to justify logical probability as being valuable for decision making, but good decision-making requires more than just coherence, if it even requires that. Hence the measurement of logical probability and the justification of an application of the theory are as yet unsolved; the former appears more likely to be settled than the latter.

Of all the theories we have considered, subjective probability holds the best position with respect to the value of probability conclusions, however arrived at. An individual who has assessed his own subjective probabilities can use them to select a decision in the expectation that his resultant choice will agree with his intuitive preferences. The user is assured of self-satisfaction. Unfortunately, the measurement problem in subjective probability is sizable and conceivably insurmountable. The individual is told roughly where to look for material out of which to fashion probabilities, but he is not told how to do so. His only guideline is that certain conclusions are prohibited. The conflict between human capabilities and the norms of subjective probability often makes the measurement of subjective probability very difficult.

2. With Respect to Definitions of Associated Concepts

The four major concepts associated with probability are random variable, independence, expectation, and conditional probability. We need say nothing about the concept of a random variable; the usual definition, as a real-valued measurable mapping, depends only on domain and range event algebras and not on probability. Independence was discussed in some detail in Section IIF where it was defined in the framework of comparative probability, in Section IIIG where the usual probabilistic definition was suggested to be a necessary but not sufficient characterization of the informal concept, and in Section VH where it was considered in the light of computational complexity and empirical tests for unrelatedness between experimental outcomes. Computational complexity provides the best viewpoint from which to study a frequency-related notion of independence. This viewpoint led us to conclude that, remarkably enough, independence could no more be adequately defined within a probability theory of the usual kind than could such notions as disjoint events. Finally there remains the possibility of a decision-oriented concept of independence as hinted at in Section IIH.

Perhaps unfortunately we have said little explicitly about expectation. If we view expectation not in its mathematical formalization as an integral of a random variable but as a measure of what it is reasonable to "expect" from a random experiment, then expectation can be given at least one natural definition in terms of comparative probability. If, as discussed in Section IIG, we can assume a payoff scaling (solvability) axiom, then we can always find a payoff $p_g \in \mathscr{P}$ that is equivalent to a gamble $g \in \mathscr{G}$ from a CP space Ω to the payoffs $\mathscr{P}$,

$$g : \Omega \to \mathscr{P},$$

where equivalence is in terms of a "rational" ordering $\gtrsim_{\mathscr{G}}$ of elements of $\mathscr{G}$;

$$(\forall \omega)\, g_p(\omega) = p, \qquad (\forall g \in \mathscr{G})(\exists p_g \in \mathscr{P})\, g \sim_{\mathscr{G}} g_{p_g} .$$

The role of expectation in utility representations of individual, rational preference patterns between decisions, as discussed in Chapter VIII, is too obvious to require further comment.

As we have suggested earlier, it is not clear that conditional probability is properly viewed as secondary to (unconditional, absolute) probability. Unlike the situation in the Kolmogorov theory, in comparative probability unconditional probability does not uniquely determine conditional probability, nor are all unconditional probability orders compatible with conditional probability orders (Section IIE). In terms of the axiomatic quantitative concept of probability, Renyi (Section IIIE) proposed a direct axiomatization of quantitative conditional probability having the advantage of being able to describe a larger family of random experiments than could Kolmogorov-type unconditional probability. Carnap, in his development of logical probability, felt it more natural to consider conditional probability, the degree to which one statement supports another, as the basic concept; statements have no absolute degree of support. Finally, in individual rational decision-making we saw that conditional probability had a role in systematizing learning from experience (Section VIIIE).

IXB. Probability in Physics

1. Introduction

It is often suggested that the (apparent) success of probability in physics is a strong argument for confidence in relative-frequency-based probability as an empirical model. To expose this illusion, we briefly examine the concept of probability in physics. Probability is used in physics primarily in the three areas of design and analysis of experiments, statistical mechanics, and quantum mechanics. The area of the design and analysis of experiments is part of statistics and its association with physics imparts no special character to it. It is the successes of the theories of statistical mechanics and quantum mechanics which are thought to be triumphs of "conventional" probability theory.

However, as we will note, the naive views of probability apparently held by physicists are a compound of classical, subjective, and finite relative-frequency elements that are strongly rejected by most who have studied inductive, inference, and decision-making, or who view themselves as thoughtful empiricists. The reason for the seeming success of these naive views is not fully understood, but it may be the manner in which the practice of physics supports the theories of physics.

2. Statistical Mechanics

Taking Tolman[†] [1] as our authority for the physicists' conception of statistical mechanics, we hear the following.

> Nevertheless, a system such as a gas composed of many molecules is actually found to exhibit perfectly definite regularities in its behavior, which we feel must be ultimately traceable to the laws of mechanics even though the detailed application of these laws defies our powers [1, p. 2].

> From a knowledge of the average behavior of the systems in a representative ensemble, appropriately chosen so as to correspond to the partial knowledge that we do have as to the initial state of the system of interest, we can then make

[†] R. C. Tolman, *The Principles of Statistical Mechanics*. London and New York: Oxford University Press (Clarendon), 1938, reprinted 1962.

predictions as to what may be expected on the average for the particular system which concerns us [1, p. 2].

We need some hypothesis, however, to guide us in choosing the probabilities and phases for different states that agree equally well with that partial knowledge. For this purpose we now introduce, as a postulate, the hypothesis of equal a priori probabilities and random a priori phases for the quantum mechanical states of a system [1, p. 350].

As is so often the case in theoretical developments, this hypothesis [see above] is not itself suitable for direct empirical test. It has, however, the indirect check that is provided by the extensive and almost astounding agreement, which is actually found between the deductive results obtained with its help and the observational results obtained on experiment [1, p. 4].

Tolman's observations suggest the following.

(1)   Uncertainty or chance phenomena in statistical mechanics arise only from ignorance of what is potentially knowable or potentially calculable.

(2)   Statistical mechanical systems exhibit regular behavior.

(3)   Probability is only an analytical tool.

(4)   It is possible to select a prior distribution that leads to close agreement between calculations and data.

(5)   The concept of probability is at best that of classical probability.

Chance only enters when there is uncertainty and uncertainty stems only from a lack of knowledge about something that is, if we wish to make the effort, knowable. The hypothesis or assumption as to the choice of equally probable cases is made only to yield a useful representative ensemble of systems. It has no other basis for belief than the agreement between calculation and observation. The existence of some hypothesis that describes the regular observations is hardly surprising. Such an analytical view of probability is unrelated to any conception we have hitherto discussed and is of no value for the analyses of inductive reasoning and rational decision-making. From the successes of statistical mechanics we gain no license to look for equally probable cases in other situations.

We can approximately account for the statistical regularity of the "random" observations which is presumably responsible for the "extensive and almost astounding agreement" found between calculation and observation in terms of the computational-complexity viewpoint developed in connection with the apparent conveyance of relative-frequency. We might well anticipate that the great complexity of most statistical mechanical systems implies statistical "regularity." Physical systems such as gases, even at very low pressures, contain at least $O(10^{10})$ particles, each having several degrees of freedom. The absence of "regular" behavior would imply a relatively low degree of complexity of the physical system and therefore an "explanation" for the anomalous behavior. Needless to say, many sources of inference and decision-making problems are not as well behaved as are the statistical mechanical systems.

3. Quantum Mechanics

Taking Feynman's well-known lecture notes[†] [2] as our authority for the meaning of probability, at least as it applies to quantum mechanics, we hear the following.

First of all, we may speak of a probability of something happening only if the occurrence is a possible outcome of some *repeatable* observation [2, p. 6-1].

By the "*probability*" of a particular outcome of an observation we mean our estimate for the most likely fraction [N_A/N] of a number [N] of repeated observations that will yield that particular outcome [2, p. 6-1].

We should emphasize that N and N_A in Eq. (6.1) are *not* intended to represent numbers based on actual observations. N_A is our best *estimate* of what *would* occur in N *imagined* observations. Probability depends, therefore, on our knowledge and on our ability to make estimates. In effect, on our common sense! Fortunately, there is a certain amount of agreement in the common sense of many things, so that different people will make the same estimate [2, p. 6-2].

The fact that one *particular* set of observations gave 16 heads most often is a *fluctuation*. We still expect that the *most likely* number of heads is 15 [2, p. 6-4].

We have defined $P(H) = \langle N_H \rangle / N$. How shall we know what to expect for N_H ? In some cases, the best we can do is to observe the number of heads obtained in large numbers of tosses. For want of anything better, we must set $\langle N_H \rangle = N_H$ (observed). (How could we expect anything else ?) We must understand, however, that in such a case a different experiment, or a different observer, might conclude that $P(H)$ was different. We would expect, however, that the various answers should agree within the deviation $\frac{1}{2}\sqrt{N}$ [if $P(H)$ is near one-half]. An experimental physicist usually says that an "experimentally determined" probability has an "error," and writes

$$P(H) = \frac{N_H}{N} \pm \frac{1}{2\sqrt{N}} \tag{6.14}$$

There is an implication in such an expression that there is a "true" or "correct" probability which could be computed if we knew enough, and that the observation may be in "error" due to a fluctuation. There is, however, no way to make such thinking logically consistent. It is probably better to realize that the probability concept is in a sense subjective, that it is always based on uncertain knowledge, and that its quantitative evaluation is subject to change as we obtain more information [2, p. 6-7].

[†] R. P. Feynman, R. Leighton, and M. Sands, *The Feynman Lectures on Physics*, Vol. I. Reading, Massachusetts: Addison-Wesley, 1963.

We do not know how to predict what would happen in a given circumstance, and we believe now that it is impossible, that the only thing that can be predicted is the probability of different events [2, p. 37-10].

These quotations suggest the following observations:

(1)   Probability has a finite relative-frequency interpretation.

(2)   There is a subjective aspect to probability that makes its actual interpretation and measurement dependent on the individual's estimates of most likely or expected values.

(3)   Individuals tend to agree in their subjective assessments of probability.

(4)   If an event A has probability $P(A)$, then in N trials we will observe $NP(A) \pm 0(\sqrt{N})$ occurrences of A.

A reliance on common-sense or right-thinking subjectivism no more clarifies the concept of probability than do references to "most likely" or "expected" outcomes. Born, who pioneered the interpretation of the square of the modulus of the Schrödinger wave function as a probability density, also came to a doubtful regard of probability as a metaphysical concept whose applicability to physics requires our belief or faith (see Section IVL). Yet a subjectivist position is only tested by its willingness to accept personal idiosyncrasies. The remarks concerning common sense suggest that the physicists only half-heartedly adopt a subjectivist position and would reject it were it really exercised.

A partial attempt to avoid subjectivism in a finite relative-frequency interpretation is evident in the introduction of confidence limits of $0(\sqrt{N})$. This device, while perhaps an accurate reconstruction of practice and deemed reasonable by Braithwaite [3], is recognized as inadequate by Feynman *et al.* We agree with the inadequacy of this device as a savior of finite relative-frequency. What then justifies the acceptance of this concept of probability? Surely not the appeal, "How could we expect anything else?"

In sum, it appears that the concept of probability fundamental to quantum mechanics, and hence to all of physics, is that of finite relative-frequency inadequately guarded by subjective common sense.

4. Conclusions

In view of the concepts of probability accepted by physicists we can only conclude that their seemingly successful applications of probability are either

(1)   irrelevant to inference and decision-making,

(2)   assured by unstated methodological practices of censoring data and selectively applying arguments,

(3)   a result of extraordinarily good fortune.

We suspect that all three points are true. The empiricist relative-frequentist with his view of probability as a limit of relative-frequency is clearly not supported by the views of the physicist; nor are any of the other positions we have discussed supported by the physicists. The work of Chaitin cited in Subsection VC6 suggests that computational complexity might be an inappropriate basis for interpreting physical randomness, for it would imply that physical systems capable of random behavior are also capable of incredibly fast computations; i.e., there is no recursive function that can bound the computation they can do as a function of time. The subjectivism of the physicists is far from that of the subjective probabilists although compatible with the thoughts of Koopman. We have indicated that the practice of physics does to some degree ensure the correctness of its theories. An example of this is provided by the quotation indicating a refusal to readjust an expected outcome of 15 heads in 30 trials even though 16 heads occurred more frequently in 100 repetitions. Finally, it is conceivable that physics is lucky in its applications of probability, at least to date. In any event, the worlds of those this book addresses do not seem to be as cooperative as the world of the physicist. Were they cooperative there would have been no incentive to develop the variety of approaches to probability which we have dicussed.

IXC. What Can We Expect from a Theory of Probability?

We can certainly ask that any proposed theory of probability be clearly formulated. Specifically, we need to know how probability assessments are initially made, how probabilities are transformed, and what the resulting probabilities can be used for. Furthermore, an interesting and rich theory will also require definitions of associated concepts that isolate important aspects of random phenomena.

With respect to the measurement or initial assessment of probabilities, we have seen that evidence or data can be either explicit (e.g., frequencies, sentences, sequences) or implicit (e.g., beliefs and opinions, unstated principles of estimation, the way we select the explicit information). Clearly both sorts of data are of importance. The difficulty is that while we can require an algorithmic specification for the conversion of explicit data into probabilities, we cannot ask that all implicit information be processed explicitly. Perhaps training procedures can be developed for the utilization of implicit information. Insofar as the conversion of explicit information is concerned, when it can be separated from implicit information, we would expect a unique prescription; for example, in the computational-complexity approach this would imply a fixed choice of universal computer. The selection of data relevant for the assessment of probability is a critical issue for a sound theory of probability and one related to the purposes of the theory. However, not much is known about this question although some discussion is available in analyses of the concept of confirmation [4].

There are several types of probability calculi. The three most commonly used structures are the modal or classificatory, comparative, and quantitative formulations. The "practical" theories have tended to gravitate to the quantitative formulation since it lends itself to relatively strong statements about inference and decision-making. Unfortunately, it is not self-evident that a useful quantitative theory is possible. Certainly if we cannot produce fully acceptable modal or comparative theories, then we cannot produce an acceptable quantitative theory. The route to the generation of a good quantitative theory may well lead through a weaker theory.

Some have argued, albeit not fully convincingly, that it is only the modal formulation that is possible. This view centers in the school of philosophers concerned with normal linguistic usage and is perhaps exemplified by the view of Toulmin [5] that probability is used to modify or guard assertions. For instance, when I tell you that "it will probably rain today" I am telling you to anticipate rain, but to be warned that it may not rain. At present, it is unclear which, if any, of these formal systems are possible ones for probability or what we can achieve with such systems.

In practice, probability is called upon to model or characterize random phenomena of chance and uncertainty and to express inferences and decisions concerning these random phenomena. With respect to the characterization of uncertainty we are faced with clarifying the relation between certainty and uncertainty. Random phenomena are often glibly characterized as those whose behavior cannot be predicted with certainty. It is assumed that we understand a primitive concept of certainty and that what is not certain, is random. There are problems both with our informal conception of certainty and with defining a class of phenomena only by saying what it is not. What, indeed, is certainty and what are examples of it? Are the only certainties logical or analytical truths? We can discern some uneasiness with "certainty" in the introduction of a notion of "practical certainty."

Viewing "uncertainty" as the opposite of "certainty" leads to the vicious circle run in by some empiricist interpretations of probability. Having started by asserting that phenomena under consideration do not have behavior predictable with certainty they then try to find something that can be said about the phenomena with certainty. While this is not a logically impossible objective, it seems to be fatal in practice. Uncertainty as the opposite of certainty seems to be a troublesome viewpoint in probability.

The other possibility is to view "uncertainty" as the primitive concept with "certainty" just an extreme case. Such a viewpoint, by only considering degrees of uncertainty, is compatible with some conclusions about random phenomena being less uncertain than others. However, to be tractable, "uncertainty" needs to be more fully explicated than it has been. Our present intuition is not very sophisticated, at least when it is verbalized.

With respect to the characterization of chance, the central question seems to be the empirical validity of probability-based predictions or statements. None of the empirical theories of probability are in reality capable of estimating probability from data alone. Facts do not speak for themselves. However, once we adjoin nonfactual hypotheses to the factual data so as to assess probabilities we run risks with the empirical validity of the conclusions. The difficulties encountered here are related to the problem of what knowledge is and are not peculiar to probability. We should not, however, expect probability theory to carry the full load of generating empirically valid conclusions. After all, probability theory is not science.

Is it reasonable to expect a single theory of probability to span the whole range of uses of probability? Looking back on our discussion of the various theories of probability we are reminded of the story of the blind men examining an elephant; each theory grasps part of the concepts of probability, chance, and uncertainty and construes it to be the whole. The existence of a unitary theory of probability is an open question [6], but not one we believe will be affirmatively answered. Furthermore, even if we could construct a unitary theory explicating the common core of the various theories of probability it is likely to be too weak to support the usual applications of probability.

The many difficulties encountered in attempts to understand and apply present-day theories of probability suggest the need for a new perspective. Conceivably, probability is not possible. A careful sifting of our intuitive expectations and requirements for a theory of probability might reveal that they are unfulfillable or even logically inconsistent. Perhaps the Gordian knot, whose strands we have been examining, is best cut. However, where would such a drastic step leave the world of practice?

IXD. Is Probability Needed?

Reflection upon the many difficulties that beset current theories of probability leads us to wonder whether we can dispense with probability and use alternative methodologies to carry out our tasks. In support of such a position is the observation that probability by itself has not been sufficient for most applications. In practice we most often make inferences, reach conclusions (estimates), or select decision rules only after adjoining an *ad hoc* statistical methodology (e.g., reduction of the class of decision rules through appeal to criteria of admissibility, linearity, unbiasedness, invariance, etc., and preferencing of rules through such criteria as minimum expected cost, minimax risk or regret, asymptotic efficiency, etc.) to a theory of probability. Why not ignore the complicated and hard to justify probability–statistics structure and proceed "directly" to those, perhaps qualitative, assumptions that characterize our source of random phenomena, the means at our disposal, and our task? This suggestion, while it has a radical element, has a long tradition to support it and examples of nonprobabilistic approaches are easily found in recent work.

The classical theory of errors [7] and, in particular, the principle of least squares [8] is a nonprobabilistic method that has long been relied on for arriving at estimates. The unknown quantity α whose value is to be inferred from observations $\{(x_i, y_i)\}$ is estimated by first modeling the source of random phenomena through a family $\{f_\alpha(x)\}$ of regression functions and then by applying the principle of least squares:

The estimate α^* of α is given by

$$\min_\alpha \sum_i (f_\alpha(x_i) - y_i)^2 = \sum_i (f_{\alpha^*}(x_i) - y_i)^2.$$

While probabilistic justifications have been given for this procedure, the procedure itself is independent of any concept of probability.

Modern theories of rational behavior [9], while often incorporating probabilistic elements, attempt to axiomatize problems dealing with decision-making in the face of uncertainty (uncertainty in the face of decision?). Axioms are provided to characterize ignorance or uncertainty as well as the notion of a rational decision procedure. The decision is made according to the hopefully unique rule that satisfies a set of axioms judged appropriate for the specific problem at hand. Examples of axiomatic rational decision-making include minimax decision-making and the Chernoff formulation of classical probability mentioned in Section VID.

The idea of fuzzy sets [10] represents an attempt at a nonprobabilistic characterization of uncertainty. Our uncertainty as to whether an x is a Y is represented by the value $f_Y(x)$ of the grade-of-membership function $f_Y(\cdot)$ for the fuzzy set Y. Whereas the usual sets of analysis are represented by a $\{0, 1\}$-valued indicator or characteristic function, fuzzy sets are described by a function that can take any value in $[0, 1]$. Unfortunately, the theory of fuzzy sets seems to have received little except a formal development; interpretive and substantive questions have been largely ignored (but see [11]).

Many of what are known as adaptive [12] or learning [13] techniques for pattern classification are essentially nonprobabilistic attempts at decision-making. The various algorithms (e.g., clustering, perceptron training, etc.) that have been proposed are nonprobabilistic in their specification and also in the thinking that led to them, although probability theory often forms part of their justification.

Recently, Kolmogorov [14] has suggested a computational-complexity-based information theory and thus a computational-complexity characterization of uncertainty. Shannon information or entropy is replaced by complexity. There has long been an interest in formulating information theory so that it is essentially independent of probability. In Wolfowitz's development of the coding theorems [15] probability appears only as an analytical tool.

As a final example of the development of nonprobabilistic techniques for treating problems usually conceived of as only analyzable through probability and statistics, we note a recent reexamination of axiomatic estimation from repeated observations [16, 17] and the axiomatic design of predictors [17, 18] capable of yielding reasonably results without recourse to statistical assumptions. Extensions of these axiomatic methods to treat decision-making appear to be possible.

Judging from the present confused status of probability theory, the time is at hand for those concerned about the characterization of chance and uncertainty and the design of inference and decision-making systems to reconsider their long-standing dependence on the traditional statistical and probabilistic methodology. Hitherto, most of the efforts at reevaluation have focused on the development of statistics without due regard to the inadequacies of the conception of probability that underlies statistics. Hence we encounter controversies between Neyman–Pearsonians and Bayesians that arise from different conceptions of probability and random phenomena and are not simply disagreements about statistical techniques. Clearly much remains to be understood about random phenomena before technology and science can be soundly and rapidly advanced. It is not only the "laws" of today that may be in error but also our whole conception of the formation and meaning of laws.

References

1. R. C. Tolman, *The Principles of Statistical Mechanics*. London and New York: Oxford Univ. Press (Clarendon), 1938, reprinted 1962.
2. R. P. Feynman, R. Leighton, and M. Sands, *The Feynman Lectures on Physics*, Vol. I. Reading, Massachusetts: Addison-Wesley, 1963.
3. R. B. Braithwaite, *Scientific Explanation*, pp. 153–186. New York: Harper, reprinted 1960.
4. C. G. Hempel, *Aspects of Scientific Explanation*. New York: Free Press, 1965.
5. S. Toulmin, *The Uses of Argument*. London and New York: Cambridge Univ. Press, 1958.
6. J. Mehlberg, Is a Unitary Approach to the Foundations of Probability Possible?, in *Current Issues in the Philosophy of Science* (H. Feigl and G. Maxwell, eds.), pp. 287–301. New York: Holt, 1961.
7. W. Smart, *Combination of Observations*. London and New York: Cambridge Univ. Press, 1958.
8. R. Deutsch, *Estimation Theory*, Chapters 4–6. Englewood-Cliffs, New Jersey: Prentice-Hall, 1965.
9. R. Luce and H. Raiffa, *Games and Decisions*, pp. 275–326. New York: Wiley, 1957.
10. L. Zadeh, Fuzzy Sets, *Information and Control* 8, 338–353, 1965.
11. R. Bellman and L. Zadeh, *Decision-Making in a Fuzzy Environment*, Rep. ERL-69-8. Berkeley: Electron. Res. Lab., Univ. of California, November 1969.
12. D. Sworder, *Optimal Adaptive Control Systems*. New York: Academic Press, 1966.
13. N. Nilsson, *Learning Machines*. New York: McGraw-Hill, 1965. ·
14. A. Kolmogorov, Logical Basis for Information Theory and Probability Theory, *IEEE Trans. Information Theory* **IT-14**, 662–664, 1968.
15. J. Wolfowitz, *Coding Theorems of Information Theory*, 2nd ed. Berlin and New York: Springer-Verlag, 1964.
16. T. Fine, A Non-Statistical Approach to Estimation from Repeated Observations, *Proc. 2nd Princeton Conf. on Information Sci. and Systems*, pp. 314–319. Princeton, New Jersey: Dept. of Elec. Eng., 1968.
17. J. Goldman, An Axiomatic Approach to Estimation and Prediction, Ph.D. Thesis. Ithaca, New York: Cornell Univ., 1970.
18. T. Fine, Extrapolation When Very Little is Known About the Source, *Information and Control* **16**, 331–359, 1970.

Author Index

Numbers in parentheses are reference numbers and indicate that an author's work is referred to although his name is not cited in the text. Numbers in italics show the page on which the complete reference is listed.

A

Aczel, J., 22(6), *56*, 73(9), *84*
Anscombe, F., 213, 216, *236*
Aumann, R., 213, 216, *236*

B

Bellman, R., 250(11), *251*
Birnbaum, A., 88(3), *116*
Black, M., 13, *14*, 113, *117*, 203, *211*
Block, H., 234, *237*
Born, M., 116, *116*, *117*
Braithwaite, R. B., 13, *14*, 245, *251*
Bruere-Dawson, G., 137, *165*

C

Carnap, R., 7, *14*, 113, *117*, 179, 186(4), 187, 194(15, 22), 195, 202(15), 203(15), *210*, *211*
Chaitin, G., 127. 129, 137, *164*, *165*
Chernoff, H., 174, *178*

D

David, F. N., 168, *177*
Davidson, D., 233(23), *237*
Debreu, G., 18(2), *56*
de Finetti, B., 87, *116*, 213, 228, 232, *236*, *237*
de Jouvenal, B., 60, *84*
Dempster, A. P., 13, *14*
Deutsch, R., 249(8), *251*
Domotor, Z., 29, 32, 35, 36, *56*
Doob, J., 76, *84*
Dynkin, E. B., 63, *84*

E

Edwards, W., 232, *237*

C (Church etc.)

Church, A., 98, *116*
Cox, R., 79, *84*
Cramér, H., 74, *84*, 90, *90*, 105(5), *116*
Cresswell, M. J., 181(6), *210*

Subject Index